T0140457

Lecture Notes in Energy 11

For further volumes:
http://www.springer.com/series/8874

Costante Mario Invernizzi

Closed Power Cycles

Thermodynamic Fundamentals and Applications

Costante Mario Invernizzi
Mechanical and Industrial Engineering
University of Brescia
Brescia
Italy

ISSN 2195-1284   ISSN 2195-1292 (electronic)
ISBN 978-1-4471-6057-1   ISBN 978-1-4471-5140-1 (eBook)
DOI 10.1007/978-1-4471-5140-1
Springer London Heidelberg New York Dordrecht

Printed on acid-free paper

Springer is part of Springer Science+Business Media (www.springer.com)

Non domandarci la formula che mondi possa aprirti,
sí qualche storta sillaba e secca come un ramo.

Eugenio Montale. Ossi di seppia, 1920–1927

Preface

The energy industry is well known for being ponderous and slow-moving, but it is undeniable that the structure of a modern society, its social, economic and cultural growth, the tenor and quality of life, in fact, the degree of civilisation and political independence of a nation or a continent, depends also on its energy potential and on the efficiency of the systems employed for its conversion and use. The electrical system of a society, for example, developed and is still developing via complicated and intricate structures that need to meet a host of needs (civil and industrial uses, a demand that varies throughout the day and according to season) and numerous available sources (oil, coal, natural gas, nuclear energy, renewable sources—solar, biomass, geothermal—the availability of reusable heat). Widely resorting to renewable energy and the distributed generation associated with cogeneration (for technological uses or for air conditioning) would introduce not only further complications but also new opportunities.

Faced with an evolving situation, most of the energy resources are consumed in combustion processes that lead to thermo-mechanical use in energy cycles and it is worth recalling certain notes made in 1984 by Prof. Gianfranco Angelino (1938–2010), regarding heat engines, where he wrote

> However varied the final destinations of the mechanical energy, there are essentially just two fundamental thermodynamic cycles: the Rankine steam cycle and the internal combustion gas cycles. The relative paucity of instruments is exacerbated further by the fact that the gas cycles are severely limited with regard to the quality of the primary energy they can use, since they are inappropriate for feeding with solid fuels or thermal energy. The Rankine cycle, while enjoying an extreme adaptability towards fuels and heat sources, is unable to use heat efficiently at temperatures above 600–650 °C, the excess heat potential being lost as driving force of the heat transmission in the primary heat exchanger. To the basic limitations of the thermodynamic concept, there are others, linked to the consolidated tradition of designing and building, whereby steam turbines are inappropriate at low power levels and alternative engines are inappropriate for large power units. In an ideal "temperature-efficiency-power unit" space, the regions satisfactorily occupied by current conversion systems are so limited in extent that there is a clear interest, not to say necessity, for researching new instruments that can expand the zone of adequate coverage along all axes

These observations are both a good introduction and a justification for the content of this book, which is specifically dedicated to closed thermodynamic cycles and their numerous applications. The need for prime movers with highly variable power levels, the use of heat sources of differing quality and composition and the search for the highest possible conversion efficiency can all find an answer in the use of heat engines based on closed thermodynamic cycles.

Closed cycles certainly present their own problems, linked (1) to the need to transfer the primary heat to the working fluid and the waste heat from the working fluid to the environment, by means of large metal surfaces for the heat exchange, then (2) to the stack losses; if the primary heat in question comes from combustion, these constitute additional losses. On the other hand, closed cycles bring undeniable advantages: (1) they have good efficiency and (2) they can use either solid fuels or the waste heat directly from chemical and industrial processes or heat with nuclear or solar origin. Furthermore, closed cycles are indispensable if the surrounding environment is unable to supply a working fluid continuously (for example, extreme cases like dynamic systems for energy production in space or in underwater systems). Then, whilst an open cycle has a fixed minimum pressure, when designing a closed cycle there are two additional freedoms: the working fluid and the minimum pressure. In principle, closed cycles offer a good solution for each specific request, restricting the dimensions needed for the turbo-machinery and the heat exchangers, reducing the level of mechanical stress in static and moving components, simplifying the dynamics and control systems and offering advantages in the off-design performances. As in the refrigeration industry, different fluids are used for different temperatures and different power levels and purposes (even though, conceptually, a gas like helium could satisfy all these); the power generation sector, likewise, could benefit greatly from the use of different working media to fit differing needs. The Organic Rankine Cycles, in this sense, are a good example of success.

In such a context, not only the thermodynamic aspects acquire relevance but also the choice of suitable materials (for instance, for high temperatures applications with non-conventional fluids) and the choice of the most appropriate working fluids for best use in the proper thermodynamic regions all become questions of primary importance. The working fluids may be pure fluids or mixtures, with well-known thermodynamic and transport properties; they must possess good thermal stability and their compatibility with other materials must be well researched, considering also the presence of any highly common impurities. The fluid dynamics of the compressors and turbo-machinery must be clearly understood in order to ensure a good aerodynamic design and high efficiency. The heat exchangers, which are often extremely demanding, require technological solutions that are not always obvious. Therefore, the study of closed cycles includes numerous aspects and represents a cross-disciplinary activity.

Chapter 1 of this book contains introductions and descriptions of the most common closed cycle heat engines: the steam cycle, in Sect. 1.6, the closed gas-turbine cycle and the Stirling engine (in Sects. 1.7 and 1.8), as important examples of closed cycle engines using an ideal gas as their working fluid. Chapter 2, forming an introduction to those that follow, discusses the thermodynamic properties of fluids, with particular regard to those aspects most closely connected to the thermodynamics of closed cycles. Organic fluid Rankine cycles are discussed in Chap. 3, while real gas closed cycles (with carbon dioxide and with organic fluids) are described in Chap. 4. Finally, Chap. 5 is dedicated to binary cycles. Among the real gas closed cycles, Sect. 4.3 looks at the real gas Stirling engines, with their peculiar and interesting characteristics. The first part of Chap. 1 summarises the thermodynamic characteristics of the heat engine, including brief references to heat pumps and refrigerating machines. The thermodynamic characteristics of the heat engine are discussed with reference to the first and second laws of thermodynamics, that is, resorting to the energetic and exergetic balances detailed in Appendix B. Meanwhile, Appendix A contains the balance equations of mass, energy and entropy. Appendix C discusses the irreversibilities in thermodynamic cycles and heat engines, with examples relative to gas cycles with perfect gas and steam cycles.

The numerous examples developed and discussed in the various chapters have been carried out with the help of the Aspen Plus® program.[1]

I believe this book will be useful for any undergraduate or post-graduate student, lecturer or junior researcher interested in applied thermodynamics and the specific aspects of energy conversion with thermodynamic engines. The basic aspects, which have been treated, I hope, in clear and rigorous fashion, could also be useful to planners in R&D Departments, who should find useful clarification in this book on fundamental theoretical elements for their specific argument.

This book would undoubtedly have been better written by Prof. Gianfranco Angelino, a thorough expert on the topic and leading figure in the Italian technology of Organic Rankine Cycles, with whom I had the opportunity to collaborate for many years. I should have liked to write the book dedicated to this subject-matter together with him, but "...around us stand the dark goddesses, one bearing the curse of sad old age and the other, death"[2] and Gianfranco died prematurely in 2010. Consequently, it has now fallen to me to write this book. My thanks are rightly due to him, though, my unforgettable maestro, first and foremost.

Next, my heartfelt thanks to my wife, Tania. Without her encouragement (and her tolerance), I should never have found the strength to conclude the work in such a relatively short time: she has been, and remains, a great help. A wonderful companion through life, along this and other arduous paths.

My thanks to Andrea Salogni, Paola Bombarda and Paolo Iora, with whom I had the opportunity to discuss to varying degrees some of the themes present here. It goes without saying that any imprecision or errors this book may contain are wholly down to me.

My thanks are also due to Andrew Neil Harwood, who helped me prepare the text in English.

Brescia, Italy
January 2013

Costante M. Invernizzi

---

[1] Aspen Plus is a registered trademark of Aspen Technology, Inc.

[2] Mimnermus (VII-VI century BC), elegy fr. 2 W, in Lirici Greci, edited by U. Albini, Le Monnier, Firenze 1972, transl. by G. Perrotta.

Contents

Acronyms

AFR	Air–fuel ratio
CCS	Carbon capture and storage
CHP	Combined heat and power
COP	Coefficient of performance
CR	Condensate recovery
EGS	Enhanced geothermal system
EPR	European pressurized reactor
FC	Fuel cell
GPU	Ground power unit
GT	Gas turbine
HP	High pressure
HTR	High temperature nuclear reactor
HVAC	Heating ventilation and air conditioning
ICE	Internal combustion engine
IGCC	Integrated gasification combined cycles
IP	Intermediate pressure
LHV	Lower heating value
LMFBR	Liquid metal fast breeder reactors
LMTD	Logarithmic mean temperature difference
LNG	Liquefied natural gas
LP	Low pressure
LPG	Liquefied petroleum gas
MCFC	Molten carbonate (K or Na) Fuel Cell
MSR	Moisture separator and reheater
MS	Moisture separator
NSSS	Nuclear steam supply system
NTU	Number of transfer unit
OCMR	Organic cooled and moderated reactors
ORC	Organic Rankine cycle
OTEC	Ocean thermal energy conversion
PAFC	(Liquid) Phosphoric acid (H_3PO_4) Fuel Cell

PCHE	Printed circuit heat exchanger
PEM	Polymer electrolyte membrane
PF	Pulverised coal-fired
RH	Reheater
SEGS	Solar energy generating systems
SCR	Selective catalytic reduction
SNAP	System for nuclear auxiliary power
SNCR	Selective non-catalytic reduction
SOFC	Solid oxide fuel cell
ST	Steam turbine
TOTEM	Total energy module
TPV	Thermo photo voltaic
USC	Ultra super critical
WHES	Waste heat recovery system

Chapter 1
The Heat Engine, the Prime Movers, and the Modern Closed Energy Conversion Systems

A heat engine is an apparatus that converts heat in useful work, usually mechanical work. It operates in a cyclical mode and makes use of a fluid: in a single-phase or in two-phase conditions. The engine receives heat from a source at high temperature, transforms a part of it into work and discharges the remaining heat fraction to the environment at a lower temperature.

The engine often consists of various machines, each one devoted to specific tasks (heat exchangers, pumps, turbines, etc.), and its operation mode can be described on the whole by means of the balance equations of energy and entropy. In the real systems, the heat available at high temperature could be the result of chemical or nuclear reactions or could be an outcome of technological processes. The balance equations of mass, energy, and entropy are summarised and discussed in Appendix A with some simple applicative examples.

The heat engine par excellence is the "reciprocating steam engine": originally designed by Denis Papin (1647–1712) [1, p. 11]. The first operating design of this kind of engine was the "Miner's Friend" version of Thomas Savery.[1] It is well known that the engine was later perfected by Thomas Newcomen.[2] Then, James Watt[3] introduced the condenser, separating it from the expansion cylinder. The studies of Rankine[4] then led, via the invention of the steam turbine in 1884, to the modern steam cycles.[5]

---

[1] Thomas Savery, 1650–1715. On 2 July 1698 Savery patented an early steam engine "for raising water by force of fire" [2].

[2] Thomas Newcomen, 1664–1729. An English ironmonger who, assisted by John Calley, a plumber, realised the "atmospheric engine." Ultimately, Newcomen's machine was capable of working with steam at low pressure, using atmospheric pressure to lower a piston in a cylinder, inside which, the mixing of water with the steam caused the latter to condense.

[3] James Watt, 1736–1819 Scottish.

[4] William John Macquom Rankine, 1820–1872. Scottish engineer and physicist.

[5] The modern steam turbine was invented by Charles Algernon Parsons, 1854–1931, an Anglo-Irish engineer.

C.M. Invernizzi, *Closed Power Cycles*, Lecture Notes in Energy 11,
DOI 10.1007/978-1-4471-5140-1_1, © Springer-Verlag London 2013

By prime movers we usually mean the machines that convert the various primary energies, which are naturally available, into mechanical energy: the steam engine first of all, then the steam turbine, diesel engine, and the gas turbine are amongst the most significant. In fact, they paved the way for the economic development of today, greatly facilitating the transport of goods and people.

A system of energy conversion normally consists of a primary mover and numerous complementary devices. A coal station, for example, requires not just the steam turbine and all the machinery that makes up the primary thermodynamic mover but also devices for heat disposal (the water circulation system, wet or dry cooling towers), for pulverising the coal (coal crushers, coal pulverisers), and its combustion (steam generators, air preheaters, etc.) and for removing the ash and treating the combustion gases before they are discharged into the atmosphere (a stack, fabric filters, electrostatic precipitators, flue gas desulfurisation systems, and SCR (selective catalytic reduction) or SNCR (selective non-catalytic reduction) reactors for the removal of nitrogen oxide).

The following sections will consider, firstly, the basic thermodynamic characteristics of heat engines, with a brief mention of refrigerating machines and heat pumps: a useful complement to thermodynamic engines, especially if combined with renewable energies and thermal storage systems. Then, consideration will be given to energy conversion systems, followed by descriptions of the traditional and principal prime movers that operate with closed thermodynamic cycles.

1.1   The Thermodynamic Characteristics of the Heat Engine[6]

In Fig. 1.1, R_H and R_C represent two heat reservoirs[7] at the temperatures T_H and T_0, respectively. The engine between the two reservoirs, operating in cyclical mode, exchanges the thermal powers $\dot{Q}_{in}$ and $\dot{Q}_{out}$ just with the two reservoirs and produces mechanical power $\dot{W}$. For the engine, we can write the power and entropy balances. In stationary conditions:

$$\dot{Q}_{in} - \dot{Q}_{out} - \dot{W} = 0 \tag{1.1a}$$

$$\dot{S}_{in} - \dot{S}_{out} + \dot{S}_G = 0 \tag{1.1b}$$

$\dot{S}_{in}$ is the entropy associated with the thermal power $\dot{Q}_{in}$, which flows from the reservoir R_H towards the engine. Likewise, $\dot{S}_{out}$ is the entropy associated with the thermal power $\dot{Q}_{out}$, flowing from the engine towards the cold reservoir R_C.

---

[6]The results in this and in the following sections can also be obtained by the rigorous application of the energy and entropy balances, as presented and discussed in Appendix B.

[7]A thermal reservoir is an ideal system which constantly maintains a stable state of equilibrium. A thermal reservoir is such in as much that any exchange of heat energy will not affect the temperature, which remains constant [3, p. 106].

Fig. 1.1 A schematic
representation of a heat
engine operating between two
heat reservoirs R_H and R_C at
temperatures T_H and T_0,
respectively

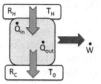

The term $\dot{S}_G$ represents the entropy generated in the engine in the time unit and
during the interaction between the engine and its surrounding systems. Thus, in the
case of Fig. 1.1:

$$\dot{S}_{in} = \frac{\dot{Q}_{in}}{T_H}$$

$$\dot{S}_{out} = \frac{\dot{Q}_{out}}{T_0}$$

and

$$\frac{\dot{Q}_{out}}{\dot{Q}_{in}} = \frac{T_0}{T_H}\frac{\dot{S}_{out}}{\dot{S}_{in}}$$

$$= \frac{T_0}{T_H}\frac{\dot{S}_{in} + \dot{S}_G}{\dot{S}_{in}} = T_0\left(\frac{1}{T_H} + \frac{\dot{S}_G}{T_H\dot{S}_{in}}\right)$$

$$= T_0\left(\frac{1}{T_H} + \frac{\dot{S}_G}{\dot{Q}_{in}}\right)$$

The efficiency whereby the engine converts the thermal power $\dot{Q}_{in}$ into the
mechanical output $\dot{W}$ is defined as

$$\eta = \frac{\dot{W}}{\dot{Q}_{in}} = 1 - \frac{\dot{Q}_{out}}{\dot{Q}_{in}} \tag{1.2}$$

Since

$$\frac{\dot{Q}_{out}}{\dot{Q}_{in}} = T_0\left(\frac{1}{T_H} + \frac{\dot{S}_G}{\dot{Q}_{in}}\right)$$

the efficiency is

$$\eta = 1 - \frac{\dot{Q}_{out}}{\dot{Q}_{in}} = 1 - T_0\left(\frac{1}{T_H} + \frac{\dot{S}_G}{\dot{Q}_{in}}\right) < 1 \tag{1.3}$$

Fig. 1.2 Discharged thermal power per unit of electrical power as a function of the net efficiency for various engines and energy conversion systems. The efficiency values are relative to the design conditions, except where explicitly indicated

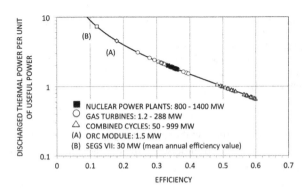

When $\dot{S}_G = 0$, the efficiency assumes the maximum value η_{max}, equal to (in the case considered in Fig. 1.1)

$$\eta_{max} = 1 - \frac{\dot{Q}_{out,min}}{\dot{Q}_{in}} = 1 - \frac{T_0}{T_H} \tag{1.4}$$

The expression (1.4), denominated the efficiency of Carnot,[8] is valid only if there are no irreversibilities in the engine or at the level of heat exchange at extreme operating temperatures. In fact, the term $\dot{S}_G$ in (1.3) takes into account the various forms of irreversibility (the thermodynamic losses, see Appendix C) that take place inside the engine and at the engine–reservoir interface: external irreversibilities (present at the edges of the system in consideration) and internal irreversibilities (inside the system). Typical irreversibility includes mechanical and fluid friction losses, heat transfer losses and throttling and mixing losses.

In any case, the thermal power $\dot{Q}_{out}$ discharged at the cold sink (usually the environment) per unit of mechanical power output $\dot{W}$ which is never zero is

$$\frac{\dot{Q}_{out}}{\dot{W}} = \frac{1}{\eta} - 1 \tag{1.5}$$

The thermal power $\dot{Q}_{out}$ is dissipated into the environment via the cooling devices of the engine and by means of any products of the combustion (exhaust gases, ashes...). High efficiency values, therefore, not only give (for equal thermal power consumed $\dot{Q}_{in}$) high useful mechanical power but also lead to reduced environmental pollution: less thermal and chemical pollution.

In Fig. 1.2 the ratio $\dot{Q}_{out}/\dot{W}$ as a function of the net efficiency for various engines and energy conversion systems is reported. An efficiency value of 50 % means

---

[8]Nicolas Leonard Sadi Carnot, 1796–1832. French physicist and engineer. Author of fundamental studies on the performance of thermal machines and considered one of the founders of the science of thermodynamics.

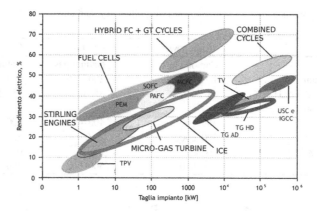

Fig. 1.3 Basic representation of the electric efficiency and of the electric unit power of the systems for energy generation (from author [5])

that the heat discharged into the environment equals the useful power. The highest values of efficiency are reached in the engines and systems with the greatest power: efficiencies of around 60 % can be obtained in combined cycles with unitary power over 500 MW. The average annual efficiency values may be significantly lower than the net nominal efficiency. For example, the SEGS power stations, steam-driven solar energy, situated at Kramer Junction in the Mojave Desert (in California), operate with thermodynamic cycles having nominal efficiency of 30–40 % and average annual values of 10–15 % [4].

Figure 1.3 shows the electrical efficiency of various systems of energy generation as a function of power. It shows the heat engines (in closed and open thermodynamic cycles) and the fuel cells (electrochemical engines). Note how a reduction in the power tends to reduce the electrical efficiency, too. By hybrid plants we mean integrated systems of fuel cells at high temperature (MCFC (molten carbonate fuel cells) and SOFC (solid oxide fuel cells)) with gas or steam cycles: these are still in the process of testing and development. If the intention is to favour distributed energy generation (that is, the large-scale diffusion of small-sized thermal engines), given the significant increase in energy performance as power is increased, there is no chance of competing with the electrical efficiency of the large combined cycles. It will be necessary to exploit to the full thermodynamic advantages of the cogeneration of heat and electricity (see Sect. 1.4). In practice, this means it is necessary to fully utilise the energy fluxes of the engine in every working condition.

Some indicative values of specific costs and efficiency with data of emission for the modern thermoelectric plants are in Table 1.1.

Table 1.1 Approximate values for investment costs, efficiency, and emissions of thermoelectric plants [6]

	Steam power plant	Gas turbine	Combined cycle
Fuel	Carbon	Natural gas	Natural gas
Specific cost (euro/kW)	1,100–1,200	200–250	400–500
Efficiency (based on LHV, %)	43–46	35–40	55–60
Specific emissions			
NO_x (g/MWh)	300–600	400–450	100–300
SO_x (g/MWh)	300–600	<10	<10
Particulate (g/MWh)	30–80	0	0
CO_2 (kg/MWh)	750–850	500–600	350–400

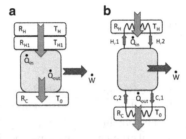

Fig. 1.4 Heat engines between a heat source at temperature T_H and a heat sink at temperature T_0 but operating at temperatures different from those of the two reservoirs. (**a**) Engine operating at maximum temperature $T_{H1} < T_H$; (**b**) engine using a working fluid that is heated from temperature T_{H1} to temperature T_{H2} ($< T_H$) and releases residual heat when cooling from T_{C1} and T_{C2} ($> T_0$)

1.1.1　Some Special Combinations of Heat Sources and Heat Engines

Let's now consider various different systems of heat engines, heat sources, and heat sinks. In particular, we discuss firstly the presence of irreversibility of heat exchange at the level of the hot reservoir (Fig. 1.4a). Then, we analyse the case in which the engine exchanges heat with the hot and with the cold reservoir using a fluid and suitable heat exchangers (Fig. 1.4b) and, lastly, the case in which the heat source is not a reservoir (at constant temperature T_H), but a fluid with variable temperature (Fig. 1.5).

If the engine (see Fig. 1.4a), which is completely reversible, receives heat from a hot reservoir at temperature T_H, but begins its conversion from a temperature $T_{H1} < T_H$, then this gives rise to an irreversibility of heat exchange equal to

$$\dot{S}_G = \dot{S}_{G,H} = \frac{\dot{Q}_{in}}{T_{H1}} - \frac{\dot{Q}_{in}}{T_H} = \dot{Q}_{in}\left(\frac{1}{T_{H1}} - \frac{1}{T_H}\right) \qquad (1.6)$$

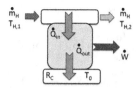

Fig. 1.5 Scheme of a heat engine receiving heat from a sensible heat source (with temperature variable between $T_{H,i}$ and $T_{H,2}$) and discharging heat into the cold reservoir at temperature T_0

with a conversion efficiency that is consequently equal to

$$\eta = 1 - \frac{T_0}{T_H} - T_0 \left(\frac{1}{T_{H1}} - \frac{1}{T_H} \right) = 1 - \frac{T_0}{T_{H1}} < 1 - \frac{T_0}{T_H} \tag{1.7}$$

If the engine is still working in a completely reversible mode but with a working fluid that receives heat from the reservoir R_H and then loses it to reservoir R_C via two heat exchangers, we get two irreversibilities linked to the heat exchange: $\dot{S}_{G,H}$ and $\dot{S}_{G,C}$ (see Fig. 1.4b):

$$\dot{S}_G = \dot{S}_{G,H} + \dot{S}_{G,C} \qquad \text{with} \tag{1.8}$$

$$\dot{S}_{G,H} = \dot{m}_H \left[(S_{H,2} - S_{H,i}) - \frac{H_{H,2} - H_{H,i}}{T_H} \right]$$

$$\dot{S}_{G,C} = \dot{m}_C \left[\frac{H_{C,1} - H_{C,2}}{T_0} - (S_{C,1} - S_{C,2}) \right]$$

and the efficiency becomes

$$\eta = 1 - \frac{T_0}{T_H} \tag{1.9}$$

$$-T_0 \frac{\dot{m}_H}{\dot{Q}_{in}} \left[(S_{H,2} - S_{H,i}) - \frac{H_{H,2} - H_{H,i}}{T_H} \right]$$

$$-T_0 \frac{\dot{m}_C}{\dot{Q}_{in}} \left[\frac{H_{C,1} - H_{C,2}}{T_0} - (S_{C,1} - S_{C,2}) \right]$$

with $\dot{m}_H$ and $\dot{m}_C$ representing the mass flows of the two working fluids in the heat exchangers.

A particularly interesting situation occurs when the engine receives heat not from a hot reservoir at temperature T_H but from a source with variable temperature (for example, from a fluid, gas or liquid, which lowers the temperature from $T_{H,1}$ to $T_{H,2}$; see Fig. 1.5). In this case, the entropy balance gives (under stationary conditions)

$$\dot{m}_H S_{H,1} - \dot{m}_H S_{H,2} - \frac{\dot{Q}_{out}}{T_0} + \dot{S}_G = 0 \qquad \text{or} \tag{1.10}$$

$$\dot{Q}_{out} = T_0 \left[\dot{m}_H (S_{H,1} - S_{H,2}) + \dot{S}_G \right]$$

and the efficiency (1.2) becomes

$$\eta = 1 - \frac{T_0}{\dot{Q}_{in}} \left[\dot{m}_H \left(S_{H,1} - S_{H,2} \right) + \dot{S}_G \right] \tag{1.11}$$

$$= 1 - T_0 \left[\frac{\dot{m}_H \left(S_{H,1} - S_{H,2} \right)}{\dot{Q}_{in}} + \frac{\dot{S}_G}{\dot{Q}_{in}} \right]$$

since

$$\dot{Q}_{in} = \dot{m}_H \left(H_{H,1} - H_{H,2} \right) \tag{1.12}$$

finally giving

$$\eta = 1 - T_0 \left(\frac{S_{H,1} - S_{H,2}}{H_{H,1} - H_{H,2}} + \frac{\dot{S}_G}{\dot{Q}_{in}} \right) = 1 - T_0 \left(\frac{1}{T_H^\star} + \frac{\dot{S}_G}{\dot{Q}_{in}} \right) \tag{1.13}$$

with $T_H^\star$ the equivalent thermodynamic temperature of the heat source and $\dot{S}_G$ representing, as usual, the sum of all the thermodynamic irreversibilities in the system being considered.

The ratio (1.13) is formally equal to the ratio (1.3) but with $T_H^\star$ in place of the temperature T_H of the hot reservoir. This latter case is representative of those situations where mechanical (or electric) energy needs to be produced in the presence of sensible heat sources. Typically: the exploitation of geothermal sources at low enthalpy and heat recovery from industrial processes (see Sect. 3.5).

1.2 The Heat Pump and the Refrigeration Machines

Figure 1.6 represents a system which, using mechanical power $\dot{W}$, has as useful effect the thermal power $\dot{Q}_{out}$ discharged at the high-temperature reservoir or the thermal power (refrigerating) $\dot{Q}_{in}$ acquired from the reservoir at temperature T_C.

The power and entropy balances, under stationary conditions, give

$$\dot{Q}_{in} - \dot{Q}_{out} + \dot{W} = 0 \tag{1.14}$$

$$\dot{S}_{in} - \dot{S}_{out} + \dot{S}_S = 0$$

Then, the entropy flows associated with the thermal powers in transit are

$$\dot{S}_{in} = \frac{\dot{Q}_{in}}{T_0}$$

$$\dot{S}_{out} = \frac{\dot{Q}_{out}}{T_H}$$

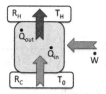

Fig. 1.6 Basic scheme of a heat pump or refrigeration cycle that uses mechanical power to produce a useful effect. The useful effect is $\dot{Q}_{out}$ in the case of a heat pump or $\dot{Q}_{in}$ in the case of a refrigerating machine

In the case of the heat pump, it is a common practice to resort to a coefficient of merit (the coefficient of performance, COP), defined as the ratio between the thermal power $\dot{Q}_{out}$ and the mechanical power consumed $\dot{W}$:

$$\text{COP}_{\text{HP}} = \frac{\dot{Q}_{out}}{\dot{W}} = \frac{1}{1 - \frac{\dot{Q}_{in}}{\dot{Q}_{out}}} \tag{1.15}$$

From the above relationships, we can derive the ratio $\dot{Q}_{in}/\dot{Q}_{out}$:

$$\frac{\dot{Q}_{in}}{\dot{Q}_{out}} = \frac{T_0}{T_H}\frac{\dot{S}_{in}}{\dot{S}_{out}} = \frac{T_0}{T_H}\frac{\dot{S}_{out} - \dot{S}_G}{\dot{S}_{out}} = T_0\left(\frac{1}{T_H} - \frac{\dot{S}_G}{T_H\dot{S}_{out}}\right)$$

or

$$\frac{\dot{Q}_{in}}{\dot{Q}_{out}} = T_0\left(\frac{1}{T_H} - \frac{\dot{S}_G}{\dot{Q}_{out}}\right)$$

and the COP of the heat pump is

$$\text{COP}_{\text{HP}} = \frac{1}{1 - T_0\left(\frac{1}{T_H} - \frac{\dot{S}_G}{\dot{Q}_{out}}\right)} \tag{1.16}$$

In the case of the refrigeration machine, the parameter of merit is calculated by the ratio between the refrigerating power $\dot{Q}_{in}$ and the mechanical power consumed $\dot{W}$:

$$\text{COP}_{\text{RM}} = \frac{\dot{Q}_{in}}{\dot{W}} = \frac{1}{\frac{\dot{Q}_{out}}{\dot{Q}_{in}} - 1} \tag{1.17}$$

Just like the previous case, the balances of energy and entropy give the ratio $\dot{Q}_{out}/\dot{Q}_{in}$:

$$\frac{\dot{Q}_{out}}{\dot{Q}_{in}} = \frac{T_H}{T_0}\frac{\dot{S}_{out}}{\dot{S}_{in}} = \frac{T_H}{T_0}\frac{\dot{S}_{in} + \dot{S}_G}{\dot{S}_{in}} = T_H\left(\frac{1}{T_0} + \frac{\dot{S}_G}{T_0\dot{S}_{in}}\right)$$

or

$$\frac{\dot{Q}_{out}}{\dot{Q}_{in}} = T_H \left(\frac{1}{T_0} + \frac{\dot{S}_G}{\dot{Q}_{in}} \right)$$

and the COP of the refrigeration machine is

$$COP_{RM} = \frac{1}{T_H \left(\frac{1}{T_0} + \frac{\dot{S}_G}{\dot{Q}_{in}} \right) - 1} \tag{1.18}$$

When $\dot{S}_G$ is null, the coefficients of performance COP_{HP} and COP_{RM} reach their maximum value:

$$COP_{HP,max} = \frac{1}{1 - \frac{T_0}{T_H}} = \frac{T_H}{T_H - T_0} \tag{1.19}$$

$$COP_{RM,max} = \frac{1}{\frac{T_H}{T_0} - 1} = \frac{T_0}{T_H - T_0} \tag{1.20}$$

A machine that absorbs a mechanical power $\dot{W}$ can function both as a heat pump and as a refrigeration machine. In the first case, the useful effect is the thermal power $\dot{Q}_{out}$ discharged at temperature T_H; in the second case, the useful effect is the thermal power (refrigerating) $\dot{Q}_{in}$ absorbed by the heat reservoir at temperature T_0. The two COPs are connected by the ratio

$$COP_{HP} = \frac{\dot{Q}_{out}}{\dot{W}} = \frac{\dot{Q}_{in} + \dot{W}}{\dot{W}} = 1 + COP_{RM} \tag{1.21}$$

Finally, the ratio (1.18) shows that it is not possible to transfer thermal power $\dot{Q}_{in}$ from a reservoir at temperature T_0 towards a reservoir at temperature $T_H > T_0$ without consuming power. The minimum power that must be consumed can be calculated by the ratio (1.20), and in the real case ($\dot{S}_G > 0$), power consumption will certainly be greater.

Exercises

1.1. We evaluate the thermal power $\dot{Q}_{out,H2}$ that can be supplied to a heat reservoir at temperature T_{H2} when there is a thermal power $\dot{Q}_{in,H1}$ provided by a heat reservoir at temperature T_{H1} and with an environment available at temperature T_0.

In the simplest and most traditional case, the thermal power available at temperature T_{H1}, usually obtained by burning fossil fuels, is transferred to the user

Fig. 1.7 Basic scheme of a
prime mover (a heat engine,
HE) and a heat pump HP for
a heating system

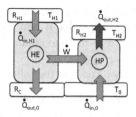

(the heat reservoir at temperature T_{H2}) via a boiler. In this case, $\dot{Q}_{out,H2} = \dot{Q}_{in,H1}\,\eta_{th}$, with $\eta_{th} < 1$ boiler efficiency.

However, referring to Fig. 1.7, we may consider a more complex and sophisticated thermodynamic system, in which the power available $\dot{Q}_{in,H1}$ is converted into mechanical power by a heat engine HE. The mechanical power $\dot{W}$ which is generated thereby is used to feed a heat pump HP which, acquiring $\dot{Q}_{in,0}$ from the environment at temperature T_0, releases the power $\dot{Q}_{out,H2}$ to the user, at temperature T_{H2}. In this case,

$$\dot{Q}_{out,H2} = \dot{Q}_{in,H1} \times \eta_{HE} \times COP_{HP}$$

with η_{HE} representing the efficiency of the engine HE and COP_{HP} which is the coefficient of performance of the heat pump HP. If T_{H1} is not too low, if T_{H2} is not excessively high, and if the system irreversibility losses are not excessive, the product $\eta_{HE}COP_{HP}$ may be significantly higher of one.

Assuming $T_{H1} = 800\,°C$, $T_{H2} = 35\,°C$ and $T_0 = 0\,°C$, and hypothesising the ideal system ($\dot{S}_G = 0$), we obtain $\eta_{HE} = 0.745$ and $COP_{HP} = 8.804$ or the ratio $\dot{Q}_{out,H2}/\dot{Q}_{in,H1} = 6.56$.

1.2. The conversion efficiency expressed by the relationship (1.3) or, more generally, by the realation (1.13) represents the fraction of the thermal power $\dot{Q}_{in}$ converted into useful power $\dot{W}$ when a heat reservoir (normally the environment) is available at temperature T_0. The efficiency η (efficiency of the first principle[9]) may vary considerably, depending on the application being considered, but it does not represent a priori a valid index of the thermodynamic quality of the conversion process. One parameter, though, that does provide useful information on the thermodynamic quality of the conversion cycle, or "good thermodynamic design" of an engine, is the second principle efficiency $\eta_{II} = \eta/\eta_{max}$.

Applying the definition,

$$\eta_{II} = \frac{\eta}{\eta_{max}} = 1 - \frac{T_0 \dot{S}_G}{\dot{Q}_{in}\eta_{max}} = 1 - \frac{T_0 \sum_j \dot{S}_{G,j}}{\dot{Q}_{in}\eta_{max}} = 1 - \sum_j \Delta\eta_{II,j} \qquad (1.22)$$

---

[9]By first principle we mean the First Principle of Thermodynamics.

with $\sum_j \Delta\eta_{\mathrm{II},j}$ representing the sum of all the losses of thermodynamic availability (see Appendix B and Appendix C) manifested in the conversion system, compared to the maximum efficiency obtainable, given a certain resource of thermal power $\dot{Q}_{\mathrm{in}}$.

We consider the following cases:

- A traditional steam vapour Rankine cycle for a power station of 300 MW, with superheating at 540 °C at a pressure of 166.7 bar and re-superheating at a pressure of 33.3 bar. Condensation temperature equal to 32.5 °C (condensation pressure of 0.05 bar). Cycle efficiency $\eta = 0.45$ [7]. The maximum efficiency is $\eta_{\max} = 1 - 305.65/813.15 = 0.624$ and the second-law efficiency $\eta_{\mathrm{II}} = \eta/\eta_{\max} = 0.45/0.624 = 0.721$.
- A steam cycle for a water-pressurised nuclear power station with water temperature exiting the reactor at 329 °C and entering at 292 °C. Efficiency $\eta = 0.34$. In this case, $\eta_{\max} \approx 1 - 305.65/583.65 = 0.476$, $\eta_{\mathrm{II}} = \eta/\eta_{\max} = 0.34/0.476 = 0.714$.
- An Organic Rankine Cycle (ORC) which recovers heat from a furnace. The heat source consists of the gases produced by combustion and available at 273 °C, which are then cooled to 130 °C. The useful electric power is 0.16 MW. The condensation temperature is 40 °C. Cycle efficiency $\eta = 0.163$ [9]. In this last case, $\eta_{\max} \approx 1 - 313.15/471.038 = 0.335$, $\eta_{\mathrm{II}} = \eta/\eta_{\max} = 0.163/0.335 = 0.486$.

The results show a substantial similarity in the thermodynamic quality of the two high-powered vapour cycles (despite the great difference in their first-law efficiency): in fact, the great power justifies the elevated plant costs, linked to the significant reduction in the irreversibility losses present in the two cycles, thereby achieving the high second-law efficiencies. The small organic fluid engine, despite having good thermodynamic qualities (equal to around 50 % of the ideal cycle), has less drive though and is less sophisticated from a plant engineering point of view.

The relationship (1.22) deserves certain considerations:

- By means of entropy analysis or calculation of the various terms $\dot{S}_{\mathrm{G},j}$, different kinds of losses (heat exchange, frictions, chemical reactions, etc.) are made homogeneous and can be compared.
- To evaluate the overall production of entropy, it is fundamental to define clearly the environment with which the system will interact (see also Appendix B and Appendix C). For instance, the preheating of the liquid in a vapour cycle may be more or less reversible according to the characteristics of the external source (a sensible heat source, like geothermal water, for example, with variable temperature, could in principle render the heat exchange reversible.
- If, in the heat transmission processes, the production of entropy is always calculated as the difference between the increase of entropy in the cold fluid and the decrease of entropy in the hot fluid, in the adiabatic transformations (expansion in a turbine, throttling in a valve), the production of entropy

coincides with the increase of entropy in the fluid which is, itself, the subject of transformation.[10]

- The entropic production (directly correlated with a thermodynamic loss) is intrinsically different from the degradation and dissipation of the mechanical work through pressure drops and friction. In fact, if $\dot{W}$ represents the mechanical power that degrades in thermal power to temperature T, the consequent entropic production is $\dot{S}_G = \dot{W}/T$, which may be significantly inferior at $\dot{W}$ if temperature T is high. The reason for this lies in the fact that the heat at temperature T has its own potential for work production quantifiable via the efficiency of an ideal thermodynamic cycle (Carnot's) between T and T_0. This potential is subtracted from the dissipation. It is only at temperature $T = T_0$ that the potential is cancelled out and the entropic loss coincides with $\dot{W}$.

From the above, we can conclude that entropic analysis reveals the definite and unrecoverable losses, where other kinds of losses may be partially recovered.[11]

1.3. The conversion efficiency that is normally intended for a system is defined by the relationship (1.2). In the case of systems using fossil fuels, the term $\dot{Q}_{in}$ is calculated as product of the flow $\dot{m}_F$ and the LHV (lower heating value) of the fuel: $\dot{Q}_{in} = \dot{m}_F \times \text{LHV}$. Other parameters traditionally referred to when using fuels are the specific fuel consumption and the specific heat consumption. Here below, we give their definitions.

- The specific fuel consumption sfc is the quantity (in mass) of fuel per unit of useful work produced. The units of measurement of the parameter sfc are usually kg/kWh :

$$\text{sfc} = \frac{\dot{m}_F}{\dot{W}} = \frac{\dot{m}_F}{\eta \dot{m}_F \text{LHV}} = \frac{1}{\eta \text{LHV}}$$

- The specific heat consumption shc (in J/kWh), on the other hand, is

$$\text{shc} = \text{sfc} \times \text{LHV} = \frac{\dot{m}_F \text{LHV}}{\dot{W}} = \frac{\dot{m}_F \text{LHV}}{\eta \dot{m}_F \text{LHV}} = \frac{1}{\eta}$$

As an example, we consider a spark ICE (internal combustion engine) with a specific fuel consumption sfc = 245 g/kWh (= 68 g/MJ). Assigning a value of LHV = 44 MJ/kg, we can determine the efficiency.

---

[10]The work $T_0 \dot{S}_{G,j}$ irredeemably lost, relative to the jth transformation, can also be calculated as the change in exergy between the start and finishing states of the process. (For the exergy calculation, see Appendix B.1).

[11]The application of entropic analysis to various typical thermodynamic cycles is discussed in Appendix C.

The efficiency can be calculated by means of

$$\eta = \frac{1}{\text{sfc} \times \text{LHV}} = \frac{1}{68 \times 10^{-6} \times 44{,}000} = 0.334$$

The two parameters, sfc and shc clearly highlight how highly efficient conversion levels correspond to lower consumption of primary energy. However, other considerations need to be kept in mind, and sometimes, these prevail in the choice and planning of a plant: limiting the weight and bulk of the system, for example, or the type of fuel used. To conclude, the conversion efficiency should be the highest possible that is compatible with the production costs of the useful energy.

1.3 The Conversion Energy Systems and the Thermodynamic Cycles

At present, 70–80 % of the consumed primary energy derives from fossil fuels (oil, natural gas, and coal). Taking just electricity production, coal accounts for around 40 % and natural gas for 20 %. In certain countries, the coal contribution to electricity generation is far higher than the world average: USA 50 %, Australia & PR China 80 % and South Africa 90 %. In Europe, Germany produces 45 % of its electricity from coal. Meanwhile, the global coal consumption in the world is rising (+7.6 % in 2010).

The consumption of natural gas is also growing rapidly in the world (+7.4 % in 2010 compared to the year before).

The conversion of the chemical energy of fossil fuels (coal and natural gas, above all) into electricity is carried out, on a large scale, in plants of hundreds of MW using steam cycles (in the case of coal) and, predominantly, combined cycles of gas turbines and steam cycles in the case of natural gas. The natural gas is also a good source of primary energy for the distributed generation of energy, thanks to its low levels of carbon dioxide and polluting emissions (sulphur dioxide, particulates, metals) per unit of energy produced (see Table 1.1). The distributed production of electricity with medium to high power plants, possibly associated with district heating, is still usually entrusted to steam thermodynamic cycles (with water vapour or organic fluids). If the useful electrical power is modest (a few kW), it then becomes interesting, from a thermodynamic and environmental point of view, to employ external combustion engines with a Stirling cycle (where possible) alongside the internal combustion engines. The thermodynamic conversion of solar energy and the use of biomass and geothermal energy can also contribute to the production of electricity, via vapour cycles (by steam or organic fluids).

To summarise, then, at least 70–80 % of the electricity currently being produced resorts to thermodynamic engines operating in closed cycles (typically, steam cycles), associated, where necessary, with open cycle gas turbines.

The conversion of heat into electricity or mechanical energy takes place via thermodynamic cycles and fluid machines which, together with all the devices and apparatus necessary for (a) treating the fuel (e.g. in the thermo-electrical or biomass stations), (b) carrying out the combustion, (c) treating and refining the products of combustion and (d) generating electricity and distributing it, all constitute the conversion energy systems.

The engines and the turbines convert thermodynamic energy (in the form of pressure and enthalpy) or kinetic energy into mechanical energy; the pumps and the compressors convert mechanical energy into pressure heads; the heat exchangers enable the transfer of thermal power between different streams of fluids.

Apart from electricity generation, other sectors using large-scale energy conversion include (a) heat production (steam generators, cogeneration systems, heat pump systems) and (b) propulsion (propellers and turbo propellers, jet propellers).

The systems for converting energy based on the prime movers are today usually realised by means of

1. Rankine cycles (with water and steam). With power typically ranging from several electrical MW up to 1,000–1,800 MW. The typical efficiencies of the great thermoelectric units are $\approx$ 45 % (with about 10 % of heat loss at the chimney stack and around 40 % at the condenser).
2. Joule–Brayton cycles (operating with a single-phase fluid). With power units ranging from a few MW to several hundreds of MW. Efficiencies are very variable ($\approx$ 20–40 %, according to the power). For the bigger machines, the typical value is 40 %, with around 58 % of the LHV (lower heating value) lost in the exhaust gases and 2 % dissipated in other ways.
3. Internal combustion reciprocating engines. From ten or so kW to ten or so MW. Efficiency varying from 20 % to 45 % according to the power. About 35 % of LHV is lost in the exhaust gases, and about the 20 % of the inlet heat is discharged as refrigerating losses.

An important distinctive characteristic of the fluid machines used in the energy conversion systems is their speed. The turbomachines normally work at speeds of hundreds of metres per second[12]; the reciprocating engines have operating speeds of tens of metres per second. Since the speed of the moving parts of the fluid machines coincide with the speed of the fluids that they contain, they are directly proportional to the fluid dynamic losses.

The usual method by which available heat is transferred to the thermodynamic engine, which, operating in a closed cycle, transforms it (in part) into mechanical or electrical work, is shown in Fig. 1.8. The heat may derive from fuel oxidation (as in Fig. 1.8) or any other origin. As the engine needs to discharge a fraction of its

---

[12]For example, the length of the LP last-stage blade of the largest nuclear turbine is today 1.75 m, with a corresponding annular exhaust area of 25.83 m^2. At a speed of 3,000 rpm, the resulting peripheral velocity is about 640 m/s. A small radial compressor (see [8]) at 250,000 rpm with a maximum diameter of 37 cm has a peripheral maximum velocity of about 480 m/s.

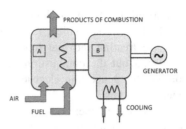

Fig. 1.8 A schematic representation of an energy conversion system with external combustion. The chemical reaction between the fuel and the oxygen of the air in the combustion chamber A releases thermal energy which, transferred to the working fluid of the thermodynamic engine B, permits the generation of electricity

thermal power into the environment, it must be cooled. In the case of combustion, the gases produced by the reaction are sent to the chimney and any solid residue (the ash) is collected and disposed of.

The Rankine cycles are always with external combustion; the Joule–Brayton cycles are cycles that usually employ internal combustion. In certain niche applications, Joule–Brayton cycles have also been designed in closed circuits with thermal power from external combustion or from nuclear reactions, solar thermal energy, etc. (see Sect. 1.7).

Cogenerative systems, as particular forms of energy conversion apparatuses are briefly described in the next section.

1.4 The Cogeneration of Thermal and Electrical Power

Cogeneration plants use the usual thermodynamic engines and are designed in such a way as to produce electricity and heat according to the two traditional schemes shown in Fig. 1.9.

A broader and more general definition for the cogeneration processes could be the following [10]:

> A cogeneration process is defined as one of various operations intended for the combined production of mechanical/electrical energy and heat, both considered useful effects, starting from any source of energy.
>
> The cogeneration process needs to make a more rational use of the primary energy than the processes that produce the two forms of energy separately. The production of mechanical/electrical energy and heat should take place mainly in interconnected sequential processes.
>
> A cogeneration system is formed by the totality of all the elements suitably combined to accomplish the cogeneration process.

The process of combined generation of electricity and heat must be "more rational" than their production separately by boiler and power station. That is, the cogeneration process must involve primary energy savings. The rational conversion

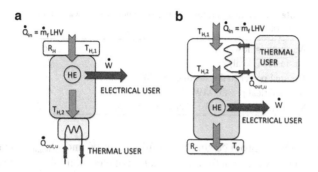

Fig. 1.9 Scheme of cogenerative systems. (**a**) With a topping engine. The engine receives heat at high temperature and the thermal utiliser collects the heat discharged by the engine; (**b**) with a bottoming engine. The thermal utiliser employs a fraction of the heat available at high temperature and the engine recuperates the residual heat, producing electricity

of primary heat energy into mechanical/electric work and useful thermal energy is obtained by cascade combination of the various machines (engines and boilers) that produce the conversion (see Fig. 1.9).

The presence of two useful effects in the cogeneration system does not give a univocal definition of its efficiency, and it is usually a common practice to define various indices of merit. For example,

$$\text{electric efficiency} \quad \eta'_{el} = \frac{\dot{W}}{\dot{Q}_{in}}$$

$$\text{thermal efficiency} \quad \eta'_{th} = \frac{\dot{Q}_{out,u}}{\dot{Q}_{in}}$$

$$\text{energy efficiency} \quad \eta' = \frac{\dot{W} + \dot{Q}_{out,u}}{\dot{Q}_{in}}$$

$$\text{efficiency of electric energy production} \quad \eta'_{el,p} = \frac{\dot{W}}{\dot{Q}_{in} - \dot{Q}_{out,u}/\eta_{th}}$$

$$\text{electric index} \quad I_{el} = \frac{\dot{W}}{\dot{Q}_{out,u}}$$

$$\text{primary energy saving index} \quad \text{ESI} = \frac{\dot{Q}^*_{in} - \dot{Q}_{in}}{\dot{Q}^*_{in}}$$

In the expression of the coefficient ESI, the term $\dot{Q}^*_{in}$ represents the primary thermal power that would be consumed producing the heat and the electricity separately in traditional plants ($\dot{Q}^*_{in} = \dot{W}/\eta_{el} + \dot{Q}_{out,u}/\eta_{th}$), whilst the term $\dot{Q}_{in}$ represents the primary thermal power (from fossil fuel) consumed in the

Table 1.2 Data and reference performances of the waste incinerator of Brescia

Waste mass flow, t/h	66
Waste LHV, kcal/kg	2,326
Thermal power to the steam cycle, MW	164
Net electrical power, MW	44
Net thermal power to the district heating, MW	110

cogeneration system to produce similar useful effects $\dot{W}$ and $\dot{Q}_{out,u}$. The efficiency η_{th}, in the efficiency expression for the production of electricity $\eta'_{el,p}$, is the efficiency of a boiler that would produce a useful thermal power equal to $\dot{Q}_{out,u}$. The adopted values of the efficiencies η_{th} and η_{el} for the power and thermal plants dedicated to the separate production of electricity and heat are usually subject to legal regulation and calculated as appropriate averages on plants with existing technology.

Exercises

1.4. The waste-to-energy facility cogeneration plant in Brescia, which started up in 2,000, exploits the combustion of solid urban waste to generate electricity and heat. It forms part of the city's district heating and by tapping the steam from the turbine, the water for the district heating can be heated. Typical operating data for cogeneration, for a winter's day and under nominal conditions, are shown in Table 1.2 [11]:

Let's evaluate the performance indices.

Assuming $\eta_{th} = 0.90$, applying the previously defined ratios and using the data in Table 1.2, we get

$$\eta'_{el} = \frac{44}{66 \times 2,326 \times 4.184 \times \frac{1}{3,600}} = \frac{44}{178.42} = 0.25$$

$$\eta'_{th} = \frac{110}{178.42} = 0.62$$

$$\eta' = \frac{44 + 110}{178.42} = 0.86$$

$$\eta'_{el,p} = \frac{44}{178.42 - \frac{110}{0.90}} = 0.78$$

$$I_{el} = \frac{44}{110} = 0.40$$

As far as the primary energy saving index is concerned,

$$\text{ESI} = 1 - \frac{\dot{Q}_{in}}{\dot{Q}^*_{in}} = 1 - \frac{\dot{Q}_{in}}{\dot{W}/\eta_{el} + \dot{Q}_{out,u}/\eta_{th}}$$

with η_{el} representing the typical production efficiency (including distribution) of the electricity. Let's assume two values for it: $\eta_{el} = 0.42$ typical for coal power stations and $\eta_{el} = 0.53$ referring to natural gas stations in combined cycle. Then,

$$ESI = 1 - \frac{178.42}{\frac{44}{0.42} + \frac{110}{0.90}} = 0.214 \qquad \text{per} \qquad \eta_{el} = 0.42$$

or

$$ESI = 1 - \frac{178.42}{\frac{44}{0.53} + \frac{110}{0.90}} = 0.131 \qquad \text{per} \qquad \eta_{el} = 0.53$$

The plant, considered in its winter set-up, always gives notable savings in primary energy (from 10 % to 20 %, according to the reference scenario) and favours the production of thermal energy: $I_{el} = 0.4$, $\eta'_{el,p} = 0.78$.

The simplified evaluations carried out here refer to powers in a single working condition; more detailed energy analysis should be referred to energy produced throughout the year.

1.5. The TOTEM, acronym for "Total Energy Module", was the first example of a small engine for cogeneration created in Italy, during the 1970s. It was designed by the Fiat Research Center and developed in Fiat Auto. The device used a 903 cm³ engine, derived from a widely used car engine and specially adapted to work with natural gas, LPG or biogas [12, 13].

Under nominal conditions, the engine operated a 15 kW alternator that supplied electricity and made available 39 kW of thermal power, heating a capacity of 3,000 L/h of water. Under nominal conditions, fuel consumption (natural gas, LHV = 35.5 MJ/Nm³) amounted to 5.7 Nm³/h.

We can calculate the performance indices, assuming $\eta_{th} = 0.9$ and $\eta_{el} = 0.53$ and with the powers at nominal conditions:

$$\eta'_{el} = \frac{15}{\frac{5.7}{3,600} \times \frac{35.5}{1,000} \times 10^6} = \frac{15}{56.208} = 0.27$$

$$\eta'_{th} = \frac{39}{56.208} = 0.69$$

$$\eta' = \frac{11 + 39}{56.208} = 0.96$$

$$\eta'_{el,p} = \frac{15}{56.208 - \frac{39}{0.90}} = 1.16$$

$$I_{el} = \frac{15}{39} = 0.38$$

$$ESI = 1 - \frac{56.208}{\frac{15}{0.53} + \frac{39}{0.90}} = 0.215$$

As shown by the results, the TOTEM system is also particularly suitable for heat production (low electrical index, modest η'_{el}, high η'_{th}, and $\eta'_{el,p}$, even greater than the unit).

In any case, it is highly unlikely that the analysis of a single parameter will allow the univocal thermodynamic evaluation of a cogeneration system. The effectiveness of the system should be evaluated by more complex procedures and energy balances, to be compared with other possible options for the plant, also bearing in mind the power levels. Another parameter for evaluating the thermodynamic quality of the process could be, for example, the second-law efficiency, which compares the useful powers $\dot{W}$ and $\dot{Q}_{out,u}$ with the maximum useful powers obtainable (electrical and thermal) for equal consumption of primary energy. Furthermore, plants generally operate with variable loads, often in ways that are difficult to predict. The various engines can use fuel of varying quality and origin, and the overall assessment must also include fundamental plant characteristics: demand for labour, reliability, maintenance, interaction with the environment (emissions, noise). Last but not least, the economic aspects (plant costs, installation, and operation) play a vital role.

1.6. If we connect a TOTEM engine, like that discussed in the previous exercise, directly to the compressor of a heat pump, the thermal user will receive heat from both the engine and the heat pump [14].

If the thermal power $\dot{Q}_{in}$ consumed in the engine is equal to 100, then 27 units are available as mechanical power which, multiplied by the COP of the heat pump, supply the thermal user with $27 \times$ COP units of thermal power, to be added to that directly recovered from the engine (69 units). Consequently, the system produces thermal energy with the relationship $\dot{Q}_{out,u}/\dot{Q}_{in} = 0.27 \times$ COP $+ 0.69$. From the energy point of view, the system is better than a traditional boiler if COP $> (\eta_{th} - 0.69)/0.27$ or, assuming $\eta_{th} = 0.9$, for values of COP > 0.78.

Figure 1.10, obtained from data reported in [16], compares various heat production technologies: direct combustion boiler, electric heat pump, several typical systems of cogeneration and district heating plants. The comparison between the various plant types is made via the mechanical equivalent of the resource consumed $\epsilon = \dot{W}/\dot{Q}_{out,u}$ per unit of heat produced as its temperature varies.[13]

The curve "a" represents the theoretical mechanical equivalent of the heat ϵ as its temperature varies:

$$\epsilon = \frac{1}{COP_{HP,max}} = 1 - \frac{T_0}{T_H}$$

with $T_C = 273$ K and T_H variable between 0 and 300°C.

Curve "a" can be associated to the straight line "b", which gives the real energy cost of heat production in the case of a boiler ($\epsilon = 1/\eta_{th}$, with $\eta_{th} = 0.85$). In the other cases considered, the coefficient ϵ was calculated as the ratio between the

---

[13] That is, the heat exergy once the temperature at which the heat is made available has been fixed. For the definition and the use of exergy balances, see Appendix B.

Fig. 1.10 Exergy cost for heat as a function of its temperature, with reference to different production technologies

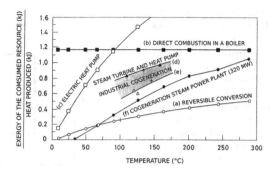

electricity obtainable (but not produced) in the plant from the heat available at the desired temperature.

The great distance separating curves "a" and "b" represents the recovery margin on which innovative heat generation technologies can impact. The performances of some of these technologies are represented by curves "c" (electric heat pump), "d" (cogeneration by means of small steam plants), "e" (cogeneration by means of great industrial plants), and "f" (cogeneration by means of great electricity production plants). All these curves give the relationship between the mechanical equivalent of the resource consumed and the quantity of heat produced. Furthermore, in a scenario that would see the dotted section (obtained by extrapolating data relative to average cogenerative technologies) accessible to most end users, the concrete value of the energy resource for direct heat uses would be drastically reduced when compared to traditional boilers.

1.5   The Traditional External Combustion Prime Movers

The current prime movers are the result of various restrictions and necessities, some thermodynamic, others mechanical or technological. As we have seen, the energy conversion system which is still most widespread is that which transforms thermal energy into mechanical by means of fluid machines,[14] so the working fluid is an important intermediary in the transformation.

The thermodynamic heat conversion comes about via an engine, inside which the fluid operates in a cycle. If the operating fluid is neither taken nor released to the surrounding environment, but always remains confined to the machine, then it is called a closed-cycle engine.

At present, the heat available is mostly produced by combustion of fossil fuels or biomass or originates from nuclear fission.

---

[14]An exception being the fuel cells, which convert the chemical energy of the fuels, usually hydrogen or methane, directly into electricity.

Fig. 1.11 Archetypal thermodynamic cycles. Both cycles have the same performance, since they work between the same temperatures and are ideal. In the Carnot cycle, the turbomachines are wholly entrusted with raising the fluid temperature; in the Ericsson cycle, this is the task of the regenerative exchanger

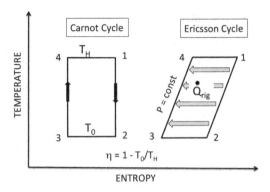

As far as the machines are concerned, the turbomachines typically have small surfaces, and the fluid passes through at high speed (hundreds of metres per second. See Footnote 12). The turbomachines are, therefore, perfectly adapted to adiabatic processes (isentropically, under the ideal conditions). The heat exchangers, with continuous flows, are appropriate for isobaric processes.

Consequently, the working fluids in the real thermodynamic cycles mainly evolve according to isobar transformations and isentropic processes. An exception, albeit an important one, are:

- The isothermal transformations in the Stirling and Ericsson engines.
- The isochoric transformations in the reciprocating internal combustion engines and the Stirling cycles (the latter being external combustion).

Then, there are constraints to the minimum temperature (typically the ambient temperature) and to the maximum temperature (the temperature of the heat sources or the maximum admissible temperature imposed by technological limits). The thermodynamic cycles, it is known, see Sect. 1.1, have a good thermodynamic quality if the available thermal power $\dot{Q}_{in}$ and the thermal power discharged into the environment $\dot{Q}_{out}$ are exchanged at extreme temperatures (e.g. T_H and T_0). In that way, the isothermal processes are favoured. Furthermore, the temperature difference $T_H - T_0$ should be (for the engines) the greatest possible, and to promote this temperature difference, the methods are:

- Using turbomachines (as in the case of steam expansion in the Rankine cycles).
- Using a regenerator (as in the ideal regenerative cycles like Joule–Brayton or the Ericsson cycles or Stirling cycles).

Figure 1.11 shows the thermodynamic temperature–entropy plane with the two ideal thermodynamic cycles that can be considered as reference models for all real cycles: the Carnot and the Ericsson cycles.[15] In general, the Carnot cycle is taken as reference for the Rankine cycles. From the point of view of thermodynamic

---

[15]John Ericsson (1803–1889) was a Swedish-American inventor and a mechanical engineer. Ericsson invented a regenerative engine in the 1820s which uses hot air. A similar engine had been

behaviour, the (ideal) Joule–Brayton regenerative gas cycle tends towards the Ericsson cycle when the compression ratio is very small.

In general, the gas cycles use air and are open cycles. If the cycle is open, the combustion can be internal and the gases produced by the combustion act directly as working fluid in the thermodynamic cycle. Closed cycles, however, have certain advantages over open ones:

- The working fluid can be specially chosen. For example, steam (water vapour) is thermodynamically better than air if the cycle is operating at relatively low temperatures.
- It is possible to employ evaporation and, especially, isothermal condensation. As these processes are isothermal, they are thermodynamically preferable.
- The phase change implies a small work of compression. As a result, even at relatively modest temperatures and with relatively inefficient machines, it is possible to obtain useful work, with good overall thermodynamic efficiency.[16]
- According to the power levels required, not only the working fluid but also the operating pressures can be purposely selected.
- The external combustion generally guarantees a greater control[17] and freedom of choice for the fuel.[18] Heat can even be used directly.

Table 1.3 lists the principal thermodynamic movers used in energy conversion, with the dates of the first working prototype.

The following sections examine the traditional thermodynamic engines being used today in closed cycles: the steam cycle (Rankine cycle), the gas cycle (Joule–Brayton), and the Stirling cycle.

---

patented in 1816 by the Reverend Robert Stirling, whose technical priority of invention provides the usual term "Stirling Engine" for the device.

[16]In gas cycles (Joule–Brayton, but also, e.g. Otto and Diesel); the differences between the compression and the expansion works are just a consequence of the average temperature of the working fluid during the two transformations (see Sect. 1.7). As a result, in the Joule–Brayton cycles, the compression and expansion operations only differ by a factor of 2–3, with the risk that the cycle will perform very badly. In the worst cases, the difference in the two works may be zero, unless the turbomachines are really very efficient. In the reciprocating internal combustion engines, since it is relatively easy to cool the engine and by virtue of its periodic operation, it is possible to reach very high maximum temperatures. So it is always possible to obtain a significant output.

[17]Inside the cylinders of a reciprocating engine, for example, the residential time interval of the gas is very short and, often, the combustion is not complete. The reciprocating engines, generally speaking, are more polluting than the external combustion ones.

[18]Gas turbines, for example, are internal combustion engines requiring combustion gases not chemically and physically aggressive to the turbine and to the combustion chamber. In other words, the gases must not be corrosive and erosive, and they should not soil the surfaces or clog up the circuits beyond well-defined and accepted limits. These constraints place restrictions on the choice of fuels.

Table 1.3 Thermodynamic energy converters and date of first working device

Steam engine	1712 (Newcomen's engine)
	1769 (Watt's engine)
Stirling engine	1816
Otto engine[a]	1876
Steam turbine	1884
Diesel engine[a]	1897[b]
Gas turbine[a]	1939[c]

[a]An internal combustion engine.

[b]On 17 February 1897, the official test gave the following results: full power 13.5 kW at 154 rpm, thermal efficiency 34.7 %, mechanical efficiency 75.5 % and net efficiency 26.2 % [15].

[c]Year of the first experimental flight of a gas-turbine military aircraft. In the same year the first gas turbine for the generation of electricity in a public power station was installed also in Neuchatel, Switzerland.

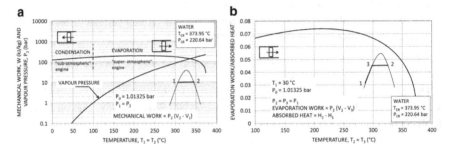

Fig. 1.12 (a) Water vapour pressure and work done during the vaporisation or condensation of a unit mass of water; (b) ratio between the work of expansion during the evaporation and the heat necessary for the steam evaporation

1.6 The Steam Cycle

The use of steam as working fluid in machines for the production of mechanical power represents the first positive attempt at industrial conversion of heat into work.

When man first seriously began to consider the possibility of substituting animal labour with the mechanical work of machines, it was discovered that most of the energy (in heat form) used for the water vaporisation was recovered as mechanical energy by great volumes of pressurised steam.

A kilogramme of water that evaporates, for example, at 200 °C, changes phase at a pressure of 15.5 bar and produces a mechanical work equal to 196 kJ/kg (see Fig. 1.12a), and the fraction of heat converted into mechanical work (see Fig. 1.12b) is around 0.07. As the evaporation temperature rises, the pressure grows rapidly until a point beyond which the mechanical work produced and the respective conversion efficiency are diminishing. This is due to the fact that the difference between the specific saturated volumes of the liquid and the steam decreases with the evaporation pressure. The condensation at subatmospheric pressures, with temperatures inferior to 100 °C, also makes very high work rates possible, for example, about 150 kJ/k with the steam condensing at 50 °C (see Fig. 1.12a).

The use of this mechanical energy by means of an engine was relatively easy, even though its conversion efficiency was very modest to start with.

Only subsequently did it become clear that the conversion efficiency could be significantly increased by making sure that the steam expansion happened in a determined way (according to an isentropic line). This led to the concept itself of a thermodynamic cycle: the Rankine cycle, which is still one of the most efficient thermodynamic engines for the generation of electricity.

1.6.1 The Thermodynamic of the Rankine cycle

The Rankine cycle is realised on the limit curve of water, and the simplest way of organising the machines (pumps, turbines, and heat exchangers) capable of creating a steam cycle is represented in Fig. 1.13a. The corresponding thermodynamic transformations for an internally reversible cycle are reported in the T–S plane (temperature–entropy) and H–S plane (enthalpy–entropy) in Fig. 1.13b, c.

Firstly, the water is taken from the minimum cycle pressure (the condensation pressure, $P_1 = P_C$) to the maximum pressure P_2, by means of one or more pumps.

Then, it is subject to heating at a constant pressure (for simplicity's sake, ignoring the pressure drops) up to the evaporation temperature $T_E = T_3 = T_4$. This temperature is that of saturation at pressure $P_2 = P_E$. Then, it vaporises between points 3 and 4; the saturated vapour thus obtained is superheated up to the maximum temperature[19] of the cycle T_5. There follows the expansion in turbine, according to the reversible adiabatic 5–6 and condensation at temperature $T_6 = T_C$.

In this way, the working fluid is returned to its initial condition, having supplied its useful effect during the expansion phase in the turbine.

Generally speaking, the following considerations apply, even for the more complex cycles:

- The condensation pressure P_C is unequivocally linked to the temperature T_C (see Fig. 1.12a), which, in turn, depends on the characteristics of the refrigerating means (cold water or air) and those of the heat exchanger (the condenser).
- The evaporation pressure P_E, unequivocally linked to the temperature T_E, also represents the maximum pressure of the cycle, the choice of which will depend on economic and technical considerations.
- The maximum temperature of cycle T_5 is chosen with regard to the maximum pressure (avoiding excessive condensation in the turbine) and on the basis of economic and technical considerations.

---

[19]If the cycle has no superheating phase, it is called saturated steam cycle. In this case $T_5 = T_4 = T_E$.

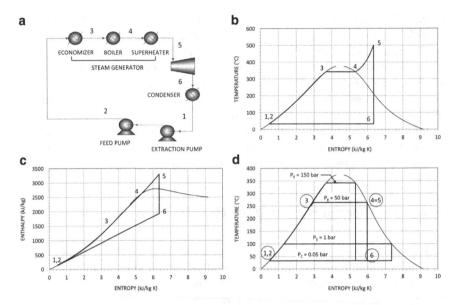

Fig. 1.13 Internally reversible Rankine cycles. (**a**) Simplified diagram of a steam plant; (**b**) thermodynamic cycle in the temperature–entropy plane; (**c**) thermodynamic cycle in the enthalpy–entropy plane; (**d**) comparison between several saturated cycles at different evaporation pressures

- The turbine and the pump make transformations which can quite reasonably be considered irreversible adiabatic.

As far as the pump compression is concerned, we observe that during an ideal adiabatic compression (isoentropic), a fluid generally undergoes a temperature increase that can be evaluated via the integration of the relationship

$$\left(\frac{\partial T}{\partial P}\right)_S = (-1)\frac{T}{C_P}\frac{1}{\rho^2}\left(\frac{\partial \rho}{\partial T}\right)_P$$

in which C_P is the specific heat at constant pressure and ρ is the fluid density. If the fluid can be considered incompressible ($\rho = const$), the heating in a reversible compression is null and its internal energy remains unchanged.

On the other hand, the heating of the liquid due to dissipation through the pump can be calculated by hypothesising that all the hydraulic losses are concentrated at the end (or at the beginning) of the compression. In this case, the temperature increase is easily calculated as the isobar heating due to the introduction into the fluid of that percentage of work lost under the form of heat

$$\Delta T = \left(1 - \eta_{\mathrm{p}}\right)\frac{W_{\mathrm{p}}}{C_P} = \frac{W_{\mathrm{p}} - \Delta P/\rho}{C_P}$$

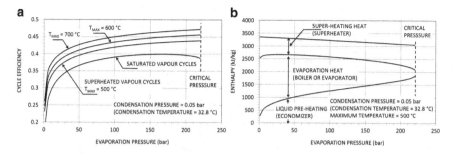

Fig. 1.14 Internally reversible Rankine cycles. (**a**) Thermodynamic cycle efficiency as a function of the evaporation pressure for saturated cycles and for superheated cycles; (**b**) heat introduced into the steam cycles at varying evaporation pressure

with η_p representing the hydraulic efficiency of the pump and W_p the compression work, which can be calculated (see Appendix A.2) as

$$W_p = \frac{\Delta P}{\rho} \frac{1}{\eta_p}$$

However, the hypothesis of incompressibility is not always acceptable (see Sect. 2.1). The close link with the pump's real operating conditions means that it must be verified each time.

Figure 1.14a reports the efficiency of internally reversible Rankine cycles, with varying evaporation pressure P_E. As the P_E increases, the efficiency increases, too, because the average temperature at which the heat is introduced gets ever higher. As the P_E grows, though, the percentage of the evaporation heat diminishes (see Fig. 1.14b), because of the shape of the limit curve (see Fig. 1.13d) and consequently, especially in the case of the saturated cycles, the improvement in the efficiency is increasingly small. The effect of the maximum temperature T_5 on the cycle efficiency is relatively modest. As T_5 rises, though, the liquid present in the steam at the end of expansion falls, greatly improving the feasibility of the cycle and ensuring high levels of expansion efficiency.

Reducing the heat evaporation percentage and increasing the evaporation pressure reduce the thermodynamic quality (the second-law efficiency) of the cycle, and the overheated cycles have a significantly lower second-law efficiency than the saturated cycles. Figure 1.15 represents the relationship η/η_{max} for saturated cycles and superheated cycles.

At the pressure of 150 bar, for example, the internally reversible saturated cycle has an efficiency of about 0.4 and a second-law efficiency of 0.80. When T_5 is 500 °C, the first-law efficiency rises to 0.43, but that of the second law becomes 0.71. With T_5 equal to 600 °C, $\eta = 0.45$ and $\eta_{II} = 0.69$. With T_5 equal to 700 °C, $\eta = 0.46$ and $\eta_{II} = 0.67$.

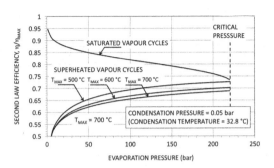

Fig. 1.15 Second-law efficiency as a function of the evaporation pressure for saturated and superheated internally reversible Rankine cycles

The relatively modest thermodynamic quality of the superheated limit cycles and the saturated cycles at high evaporation pressure is due to the irreversibilities $\dot{S}_{G,j}/\dot{Q}_{in}$ present in the various phases of heat transmission between the heat source at temperature T_5 and the working fluid in the cycle. With reference to Fig. 1.13,

$$\frac{\dot{S}_{G,2-3}}{\dot{Q}_{in}} = \frac{S_3 - S_2}{H_5 - H_2} - \frac{H_3 - H_2}{(H_5 - H_2)\,T_5} \qquad \text{heat transfer irreversibility in the economiser}$$

$$\frac{\dot{S}_{G,3-4}}{\dot{Q}_{in}} = \frac{S_4 - S_3}{H_5 - H_2} - \frac{H_4 - H_3}{(H_5 - H_2)\,T_5} \qquad \text{heat transfer irreversibility in the boiler}$$

$$\frac{\dot{S}_{G,4-5}}{\dot{Q}_{in}} = \frac{S_5 - S_4}{H_5 - H_2} - \frac{H_5 - H_4}{(H_5 - H_2)\,T_5} \qquad \text{heat transfer irreversibility in the superheater}$$

For example, at pressure $P_E = 150$ bar and $T_5 = 600\,°C$, the irreversibility linked to the preheating of the working fluid during phases 2–3 contributes to total thermodynamic losses for 68 %, the evaporation phases 3–4 contributes 21 % and the superheating 4–5 the remaining 11 %.

On the downside, the high value of $\eta_{II} = \eta/\eta_{max}$, in the case of saturated cycles at low temperature (see Fig. 1.15), is the consequence of the high ratio between the heat of evaporation, introduced at a constant temperature equal to that of the hot reservoir, and the heat of the preheating.

Having established the maximum and minimum temperatures, then the increase in the thermodynamic efficiency of the Rankine cycle must pass through a reduction of the irreversibility (external to the cycle) of the heat exchange during the heating phase of the liquid, along the isobars 2–3.

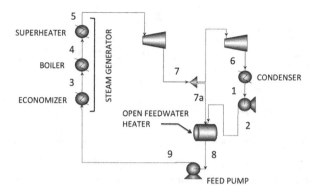

Fig. 1.16 Simplified scheme of steam cycle with one vapour extraction for regeneration

1.6.2   The Regenerative Rankine Cycle

As we have seen, heating the liquid in a simple thermodynamic steam cycle, as shown in Fig. 1.13, leads to a significant loss in efficiency, as it takes place under great temperature differences in the boiler. In particular, the preheating in the liquid section, from point 2 to point 3, which happens in the presence of the greatest temperature differences between the fluid and the heat source, is responsible for the largest thermodynamic loss. A substantial reduction in this heat exchange loss can be obtained by preheating, by using percentages of steam that are gradually extracted at ever lower temperatures as the vapour expands in the turbine.

Figure 1.16 shows a cycle with the steam extracted just once, at a pressure P_7, midway between the maximum pressure (the evaporation pressure $P_E = P_5$) and the minimum pressure (the condensation pressure $P_C = P_6$). The percentage of steam that is extracted is mixed in the open feedwater heater with the part of liquid that comes from the extraction pump of the condensate. The liquid originating from the condenser is, thereby, heated from temperature T_2 to temperature T_8 at the expense of the latent heat of condensation of that percentage of steam extracted in 7. The two flows, once they are mixed (at the thermodynamic conditions of point 8), proceed towards the steam generator by means of the feed pump.

Figure 1.17a reports the efficiency of the thermodynamic cycle of Fig. 1.16 as a function of the steam extraction pressure. At the optimum value of the pressure P_7, the heat exchange losses (the only ones present in the internally reversible Rankine cycle considered here) are minimal: this optimal pressure value of P_7 corresponds with the maximum efficiency (see Fig. 1.17b).

In the simplified ideal case under consideration, the thermodynamic efficiency of the cycle passes from a value of 0.45 to 0.48: a net gain of three points. The open feedwater heating introduces a mixing loss, which is absent when there is no steam extraction. However, the significant reduction in irreversibility in the next phase

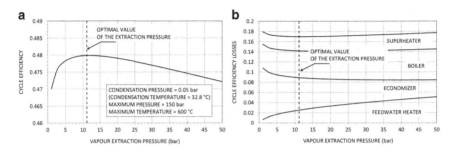

Fig. 1.17 Internally reversible Rankine cycles. (**a**) Cycle efficiency at varying pressure of extracting the steam; (**b**) loss of cycle efficiency due to the irreversibility of plant components

of heating the liquid amply compensates for the minor irreversibility linked to the mixing.

The drops in efficiency $\Delta\eta_j = T_0\dot{S}_{G,j}/\dot{Q}_{in}$, reported in Fig. 1.17b, with $j = 1, 2, 3, 4$ and $T_0 = T_1$, can be calculated using the following relationships:

$$\frac{\dot{S}_{G,1}}{\dot{Q}_{in}} = \frac{\dot{m}_8 S_8 - \dot{m}_2 S_2 - \dot{m}_{7a} S_7}{\dot{m}_8 (H_5 - H_9)} \qquad \text{mixing irreversibility in the feedwater heater}$$

$$\frac{\dot{S}_{G,2}}{\dot{Q}_{in}} = \frac{S_3 - S_9}{H_5 - H_9} - \frac{H_3 - H_9}{(H_5 - H_9) T_5} \qquad \text{heat transfer irreversibility in the economiser}$$

$$\frac{\dot{S}_{G,3}}{\dot{Q}_{in}} = \frac{S_4 - S_3}{H_5 - H_9} - \frac{H_4 - H_3}{(H_5 - H_9) T_5} \qquad \text{heat transfer irreversibility in the boiler}$$

$$\frac{\dot{S}_{G,4}}{\dot{Q}_{in}} = \frac{S_5 - S_4}{H_5 - H_9} - \frac{H_5 - H_4}{(H_5 - H_9) T_5} \qquad \text{heat transfer irreversibility in the superheater}$$

with $\dot{m}_2 + \dot{m}_{7a} = \dot{m}_8 = \dot{m}_7$.

The mixing irreversibility grows with the rising extraction pressure, whilst the irreversibility in the preheater falls rapidly. The thermodynamic losses in the boiler and the superheater remain practically steady with the pressure of the steam extraction. The rapid fall in the losses in the economiser, together with the increased losses through mixing, is responsible for the minimum value of $\Delta\eta = \sum\Delta\eta_j$ corresponding, in our example, to an extraction pressure of around 11 bar.

In general, the greater the positive effect of the regeneration, the greater the maximum pressure of the cycle and, in practice, in modern USC (Ultra Super Critical) steam plants dedicated to electricity production, the steam extractions are at least ten or so.

The extracted steam is brought into thermal contact with the feedwater by means of surface exchangers or, more rarely, by mixing. Given the excellent exchange characteristics of both the condensing steam and the water, the differences in

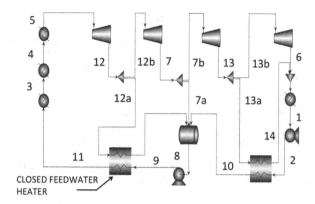

Fig. 1.18 Simplified scheme of a steam cycle with three steam extractions for regeneration

end temperature, when exiting the exchanger, are generally very small (3–5 °C). Figure 1.18 represents a cycle with two closed feedwater heaters and one intermediate open feedwater heater.

The basic criterion for establishing the route of the condensed steam at the exit of each exchanger is still that of reducing the irreversibility as much as possible or, what amounts to the same thing, the temperature difference between the different fluids in thermal contact. This explains the passage of the condensate in cascade from one exchanger to that immediately before (see Fig. 1.18).

The mixers, by consenting null differences in end temperatures, have a higher thermodynamic efficiency. However, they oblige the plant to undergo a further imposition, namely, each exchanger must be fed by a pump, which applies the same pressure of the extracted steam to the feedwater. The use of surface exchangers, though, means that the number of pumps can be drastically reduced. Commonly, at least one mixer is always present to carry out the role of a deaerating heater.[20]

Once the general layout of the cycle and the number of regenerators have been decided, there remains the question of determining the optimal temperature for the end of regeneration and the distribution of the feedwater enthalpy increase among the various exchangers.

---

[20]The water, which constitutes the working fluid for the power stations, must be not only demineralised but also non-aerated. In fact, at high temperatures, the oxygen from the air which has dissolved in the water becomes corrosive for iron materials. This degassing process is carried out with a special device, the deareator, consisting of a mixture heat exchanger in which the liquid, shaped as thin sheets and free-falling, exposes to the gas-phase large surfaces for heat and mass exchange. The dissolved gas, in first approximation, even within the liquid, obeys the law of perfect gases and, in the absence of atmospheric air, expands (abandoning the liquid) and tends to assume a specific volume that corresponds to the partial residual pressure of the air in the exchanger. This pressure is kept low, driving away the air as it is gradually separated. Although, in principle, the degassing is possible at any temperature, experience teaches that it is most efficient at 100–150 °C.

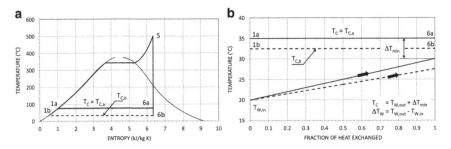

Fig. 1.19 Effects of a decrease of the condensation temperature on a Rankine steam cycle. (**a**) Thermodynamic cycle in the T–S plane. (**b**) Diagram of heat exchange at the condenser

The definite resolution of the problem is not simple as the final result depends on a great number of variables, such as the cycle layout (number of superheatings, the pressure, and temperature at which they take place), the efficiency of the rotating machines and the static equipment and the thermodynamic cycle of the gases produced by the combustion (in the case of power plants, using fossil fuels). The basic concept remains that of the overall reduction of irreversibility, as achieved in the simple case of the system in Fig. 1.16, with the results shown in Fig. 1.17.

It is not uncommon for structural and economic problems to overshadow the strictly thermodynamic aspects (e.g. the choice of end temperature for the regeneration could depend upon economic considerations, bearing in mind, for example, that the cost of the steam generator strictly depends on the degree of regeneration too). The final decision on the number of regenerators, therefore, depends on optimising the unit cost of the energy produced.

1.6.3   The Importance of the Condensation Pressure

The pressure at the end of expansion (the condensation pressure) has a notable influence on the thermodynamic performance and characteristics of the steam cycle.

For example, [47] referring to Fig. 1.19a, if the expansion progresses from point 6a to point 6b, that is, the condensation temperature passes from $T_{C,a} = T_C$ to $T_{C,b}$, the heat introduced into the cycle grows by a quantity $\Delta Q_{in} = (H_{1a} - H_{1b}) \approx C_P (T_{1a} - T_{1b}) = C_P (T_{C,a} - T_{C,b}) = C_P \Delta T_C$, whilst the useful work increases by a term ΔW equal to the area $(6b - 1b - 1a - 6a)$:

$$\Delta W \approx (T_{C,a} - T_{C,b}) (S_{6b} - S_{1b}) \approx \Delta T_C \frac{\Delta H_{VL}}{T_C} x_{6b}$$

with x_{6b} representing the vapour quality of the point 6b.

The drop in the condensation temperature has a positive effect on the thermodynamic performance of the cycle, if the relationship

$$\frac{\Delta W}{\Delta Q_{\text{in}}} = \Delta T_{\text{C}} \frac{\Delta H_{\text{VL}}}{T_{\text{C}}} x_{6b} \frac{1}{C_{\text{P}} \Delta T_{\text{C}}} \approx x_{6a} \frac{\Delta H_{\text{VL}}}{C_{\text{P}} T_{\text{C}}}$$

assumes values greater than one unit. Assuming a reference condensation pressure equal to 0.05 bar and a vapour quality x_{6a} of about 0.9, we obtain

$$\frac{\Delta W}{\Delta Q_{\text{in}}} \approx x_{6a} \frac{\Delta H_{\text{VL}}}{C_{\text{P}} T_{\text{C}}} = 0.9 \frac{2,424 \times 10^3}{4.2 \times 10^3 \times 305.55} = 1.7$$

Therefore, the original cycle undoubtedly benefits thermodynamically from reducing the condensation temperature. The actual convenience of the procedure, or better the optimisation of the condensation temperature, is subordinate though to calculating the unit cost of the electricity produced, as a function of the condensation temperature.

To that end, refer to Fig. 1.19b, where we report the diagram of heat exchanged in the condensation phase of a steam Rankine cycle. The temperature T_{C} is set on the basis of the temperature of the refrigerating water $T_{\text{W,in}}$, on the basis of temperature increase $\Delta T_W = T_{\text{W,out}} - T_{\text{W,in}}$ and on the basis of the minimal temperature difference ΔT_{min} of the two streams in thermal contact. A lowering of the condensation temperature, for a given temperature of the refrigerating water, brings about either a drop in ΔT_{W} (reflected in a greater flow of refrigerating water) or a drop in ΔT_{min} (reflected in an increase of the exchange surface in the condenser) or in both of these.

With great quantities of cold water available, it is possible to push the condensation pressure down very low. For example, 0.025 bar for $T_{\text{W,in}} = 6\,°\text{C}$, $\Delta T_{\text{W}} = 8\,°\text{C}$, $\Delta T_{\text{min}} = 7\,°\text{C}$ and $T_{\text{C}} = 21\,°\text{C}$.

In conclusion, whilst the cycle certainly benefits thermodynamically from the drop in condensation temperature, the following factors have a negative impact:

- The increase in the exchange surface and, consequently, in the cost of the condenser
- The increase in the power and, therefore, in the fixed and operating costs, in the circulating pumps for the cooling water
- The increase in capacity and, thereby, the cost of the extractor for the non-condensing gases in the condenser
- The increase in the dimensions of the exhaust ring on the turbine

This last item is often the most costly. In fact, the volumetric capacity, in passing from a condensation pressure of 0.05 bar to a pressure of 0.025 bar, increases by 1.8–1.9 times. In a turbo-alternator with limited power (50–100 MW), for example, this may mean substituting the single-flow solution at low pressure with a double-flow alternative, making the machine considerably more expensive.

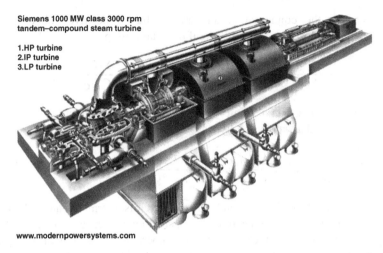

**Siemens 1000 MW class 3000 rpm
tandem–compound steam turbine**

1.HP turbine
2.IP turbine
3.LP turbine

www.modernpowersystems.com

Fig. 1.20 Tandem-compound steam turbine and the shell and tube condenser below the turbine. 1,000 MW class turbine at 3,000 rpm (from author [17])

Increasing the power of the pumps for the refrigerating water and of the extraction devices for the air from the condenser further impacts, in the real case, the reduction in fuel consumption that was calculated in the ideal case. Finally, reducing the exhaust pressure of the turbine increases the percentage of condensate, with fluid dynamic and technological damages (passing from a condensation pressure of 0.05–0.025 bar, the vapour quality drops, for instance, from 0.90 to 0.88). The optimal value for ΔT_{min} is about 5 °C, and, at least in the Italian climate, an average annual value of the condensation temperature of 32 °C is considered a realistic reference: this corresponds to a saturation pressure of 0.05 bar.

Figure 1.20 represents a modern steam turbine for a modern thermoelectric station (that of Niederaussem K, Germany) in the tandem-compound configuration, with a single-flow HP (high-pressure) section, with a double-flow IP (intermediate pressure) cylinder and with three double-flow LP (low pressure) cylinders for the condensing section, on a full speed single shaft at 3,000 rpm. The shell and tube condenser is located beside the exhaust sections of the LP section.

Since the quantity of heat that needs to be released into the environment may be greatly superior to the useful work produced (see (1.5) and Fig. 1.2), the quantity of water necessary for the condensation may be remarkable (e.g. 60 times the flow of steam to the condenser; see Problem 1.7), and such a quantity of water is not always available. Furthermore, given the great thermal power involved, there may be a significant thermal alteration in the natural water bodies, with harmful effects on the ecosystem (of a physical, chemical, and biological nature). In many countries, there are regulations governing the discharge of hot water into bodies of natural water, with the express intent of limiting excessive increases in water temperature and preventing the formation of "thermal barriers." Consequently, alternative cooling systems are resorted to, and the system currently most in favour is that of cooling towers.

Fig. 1.21 Calder Hall power station. Calder Hall had four natural draught cooling towers, each one 88 m high, built between 1950 and 1956 (Source Magnox Electric Ltd)

The wet cooling towers or the evaporative cooling towers, for the most part natural draught cooling towers, are usually made of great hyperbolic shells, standing vertically, with openings at both the top and the bottom and with a shallow permeable layer at the bottom end (see Figs. 1.21 and 1.22). The hot water, on leaving the condensers, is sprayed from above onto this layer, creating a wide contact surface with the air, in order to exchange mass and heat. The saturated hot column of air (which is lighter for both these reasons), which emerges from this layer, rises upwards, thanks to the chimney effect of the shell, summoning more air, which is relatively dry, from the outside through the openings at the base of the tower. The transfer of the latent heat of vaporisation through the saturation of the air cools the water, which is collected at the bottom and sent once more to the condensers. Water consumption is limited, but not unimportant (around 2 kg per electrical kWh).

Where there is no adequate natural availability, a further option is the direct exchange of heat with the atmosphere by means of devices called dry cooling towers. These consist of heat exchangers with a large surface on the air side and a tubular primary surface, inside which there flows either, directly, the vapour to be condensed (direct dry cooling condensers) or, indirectly, as in the indirect dry cooling tower, a flow of water for the purpose of cooling the condensers.

The very modest thermal capacity of air implies the use of enormous flows, whilst the small external exchange coefficients require the use of vast surfaces. The problems of the heat exchange are similar to those found in car radiators.

Whilst the evaporative towers incur a limited cost increase in the energy generated (greater investment and an increase in the condensation temperature, albeit limited), the dry towers imply significant extra costs due not only to the further

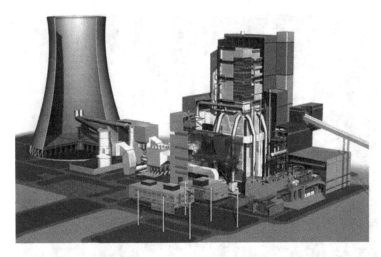

Fig. 1.22 Carbon capture and storage (CCS) pilot plant in Niederaussem, Germany. *On the right*: lignite conveyor, turbine building and steam generator, the double gas desulphurisation system and the cooling tower. The cooling tower serves also as a chimney for the units. The cooling tower is with natural air draught (from author [18])

Fig. 1.23 Detail of an air-cooled condenser of the combined power plant of Gissi (Chieti, Abruzzo, Italy). Two units of 400 MW each one. In the figure, the ducts of the steam and the fans for the cooling air are visible

investment but also to the not-insignificant rise in the condensation temperature and the energy consumption for the fans that guarantee the flow of air through the heat exchanger.

Figure 1.23 shows an air condenser, in which the exhausted steam from the turbines is cooled directly by air.

Fig. 1.24 The inlet turbine temperature for ideal steam cycles ensuring a vapour quality at the end of the expansion equal to 0.85 as a function of the evaporation pressure

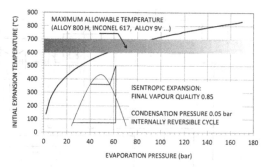

The use of cooling towers resolves partially (if wet) or completely (if dry) the problem of consumption and the excessive heating of natural bodies of water, but it does present new ones: the alteration of the local microclimate and chemical pollution (for wet towers) as well as noise pollution for the dry towers. Furthermore, the plant becomes more sensitive to changes in the environmental conditions, and during the course of the year, the discharge pressure of the turbine may vary considerably. In conclusion, the choice between conventional systems and cooling towers should be based on an overall economic evaluation that also takes into consideration the social costs due to pollution and not just the simple running costs of the power station.

1.6.4   The Superheating and the Repeated Superheating

Almost all the steam cycles foresee superheating, the main function of which is to increase the vapour quality at the end of expansion. In fact, an excessive liquid content causes erosion of the blades in the turbine. Qualities of 0.85–0.95 are considered at the limit of acceptability. A second reason for favouring superheating is the increased internal efficiency of the turbine stages as the quality increases: on the basis of a simple empirical rule, the efficiency of a stage that works with wet steam is equal to that in the superheated region multiplied by the quality. Lastly, there is a third influence that favours superheating: the thermodynamic improvement caused by introducing added heat at high temperature.

Historically, since the middle of the last century, maximum cycle pressures have shown a significant upward tendency, pushed by technological progress and the increased efficiencies this made possible. At pressures of 120–180 bar and for start of expansion temperatures of 540–600 °C, the superheating does not guarantee a sufficient quality at the end of expansion. Figure 1.24, for example, shows, as a function of the vaporisation pressure, the superheating temperature necessary to maintain the quality at the end of an isentropic expansion equal to 0.85. As shown, the superheating temperature needed grows rapidly with the evaporation pressure.

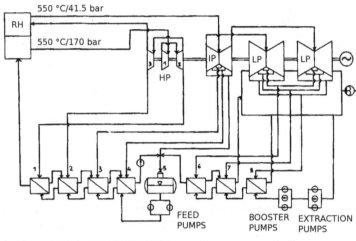

PIACENZA LEVANTE 320 MW (1967)

Fig. 1.25 The simplified scheme of a real thermoelectric steam cycle. The Piacenza–Levante Power Plant (Italy), one of the two similar conventional units erected between 1965 and 1967 in Italy. Before the revamping in 2005 with a modern gas-turbine combined cycle

In principle, then, the use of higher temperatures (700–800 °C) could bring the ratio to an acceptable level, but such temperatures mean using sophisticated materials, especially for the steam generator where the amount of metal used is significant. Consequently, for economic reasons rather than technical ones, there has been no move to further increase the maximum temperature. As an alternative, the current practice is to resort to a second superheating, after a first expansion stage in the turbine. The entire steam flow is extracted downstream from a high-pressure group and sent to the superheater in the steam generator, from where it returns to the turbine room at maximum temperature. The benefits of resuperheating are identical to those of a simple superheating.

The double transfer of steam from the turbine to the boiler and back again, plus the added complexity required in the turbine and boiler for the resuperheating phase, makes this practice relatively expensive: nevertheless, the increased cost is amply compensated by the improved performances of the cycle. In certain cases, the standard plant scheme is replaced by a variant (typically, e.g. in steam cycles in nuclear power plants, see Sect. 1.6.6) which involves in direct superheating of the expanded steam by means of the live steam produced by the boiler; naturally, given the differences in temperature needed to enable the heat exchange, the resuperheating temperature, in this case, will be somewhat lower than that of the live steam.

Figure 1.25 shows a simplified drawing of a steam cycle (the old power station of Piacenza Levante, Italy) with superheating at 550 °C, followed by a resuperheating at the same temperature, but with an intermediate pressure of 41.5 bar. The steam

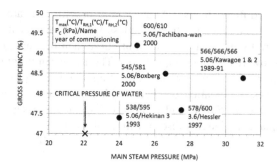

Fig. 1.26 Performances of supercritical steam plants. The rated outputs of the plants are from 700 to 1,000 MW. In the figure, T_{max} indicates the maximum steam temperature; $T_{RH,1}$ and $T_{RH,2}$ are, respectively, the superheating temperature and the resuperheating temperature (where relevant). P_C indicates the turbine back pressure (the condensation pressure)

is extracted eight times for the regeneration. Extraction number 5 concerns the deaerator. Given the high power level, the boiler feed pumps are often dragged by an auxiliary turbine, fed by steam that is tapped when required, usually at the same pressure as the deaerator.

1.6.5 Supercritical Cycles

The cycles that we have discussed so far have involved the isothermal vaporisation of a preheated liquid. As technology has progressed, it has become possible to create cycles with pressures greater than critical (221.3 bar) in which the isothermal vaporisation is replaced by transition to a temperature that is constantly rising, from a liquid state to a vapour one. In Fig. 1.26, from data reported in [19], we can see the gross efficiency as a function of the main steam pressure for various supercritical steam plants. Together with supercritical cycles at pressures of 250–300 bar, it has become technically and economically feasible to propose double resuperheating. However, those power stations that are capable of handling it are still in a minority. Designing the resuperheating section becomes complicated, adding a second turbine section at high temperature and, above all, bringing in additional steam-lines that have a significant diameter. For these reasons, bearing in mind the relatively modest increase in efficiency (around 1 %), those plants with maximum pressures below 300 bar (30 MPa) have preferred not to adopt a second superheating. The maximum efficiencies obtainable for temperatures lower than 600 °C stand at around 47–48 %, corresponding to a plant efficiency of 42–43 % (steam generator efficiency about 0.9). Although there have been cases of temperatures of 650 °C being used, in general, temperatures of no more than 540–560 °C are favoured because this minimises the use of steel superalloys which are not only extremely expensive but also difficult to use. After all, the benefits that are derived from increasing the temperature are relatively modest, since the barycentre of the cycle is determined more by the shape of the limit curve than the further extension of the superheating towards higher temperatures. The European programme AD700 ("Advanced supercritical PF power plant operating at 700 °C" [20]), in existence since 1998, foresees the creation of cycles with a maximum steam pressure of 350 bar, superheating of 700 °C and resuperheating (at a pressure of 75 bar) of 720 °C. The net plant efficiency foreseen, with cooling towers, is 50–52 %.

1.6.6 The Steam Cycle for the Nuclear Power Plants

In the nuclear reactors moderated and cooled by light water, pressurised or boiling, which currently form the great majority of nuclear power plants operating today, the steam originating from the "Nuclear Steam Supply Systems" (NSSS) has a pressure of around 70–80 bar and is saturated (temperatures of 290–320 °C). The thermodynamic characteristics of the steam under these conditions give rise to a series of problems, the solution of which makes the steam cycle for nuclear power plants special.

The relatively modest enthalpy drop available (e.g. 1,000 kJ/kg, compared to 1,500 kJ/kg in the case of steam at 500–600 °C and 180 bar) and the lower thermodynamic efficiencies imply the use of correspondingly much higher flow masses (at similar horsepower), which, given the usual condensation pressures, gives rise to exceptionally high flow volumes at the discharge of turbine bodies operating at low pressure. For this reason, the nuclear turbines always have a rotating speed of 1,500 rpm (in the 50 Hz systems) or 1,800 rpm (in the 60 Hz systems), which is half the rpm usually adopted in the more traditional thermodynamic steam cycles.

By starting expansion from near-saturation conditions, the humidity in the steam grows rapidly, creating a potentially intolerable erosion of the rotor blades in the turbine and reducing its fluid dynamics performances, too. There should always be a moisture separator for the entire steam flow, generally positioned between the bodies at high and intermediate pressure, followed by a slight intermediate superheating with live steam. Figure 1.27 represents a basic layout for drying the steam and the resuperheating in two stages in a block called MSR (Moisture Separator Reheater). This layout is that adopted by the power station of Flamanville 3 [21].

The qualitative diagram of heat exchange in the superheater RH of the block MSR is shown in Fig. 1.28. The steam with thermodynamic conditions of point C is superheated by means of two extractions A and B at two different pressures. The steam A is a portion of the live steam derived from the NSSS.

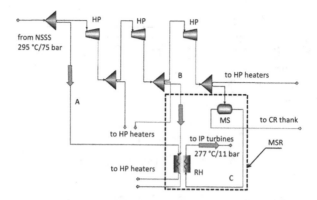

Fig. 1.27 Basic layout of the drying system and steam superheating for a turbine of a nuclear reactor

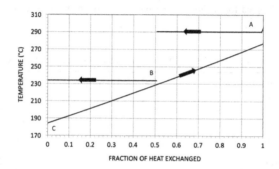

Fig. 1.28 Diagram of the heat exchange in the superheater RH of the MSR system from this figure

Exercises

1.7. Figure 1.29 shows the plant layout (simplified) of a thermoelectric plant of 320 MW (the old power station of Piacenza–Levante, Italy; see Fig. 1.25), and Table 1.4 has the basic parameters adopted for the cycle calculations.

The regenerators H1 and H2 are fed with two steam extractions from the HP stages of the turbine. The surface regenerators H3 and H4 and the deareator H5 are fed by steam extractions from the IP stages of the turbine. The regenerators H6, H7, and H8 use steam extracted from the LP turbine.

The cycle efficiency (excluding that of the boiler) is calculated as the ratio between the useful horsepower and the thermal power entering the cycle.

$$\eta = \frac{\left(\sum \dot{m}_{\mathrm{HP}i}\,W_{\mathrm{HP}i} + \sum \dot{m}_{\mathrm{IP}i}\,W_{\mathrm{IP}i} + \sum \dot{m}_{\mathrm{LP}i}\,W_{\mathrm{LP}i}\right) - \left(\dot{m}_{\mathrm{FP}}W_{\mathrm{FP}} + \dot{m}_{\mathrm{BP}}W_{\mathrm{BP}} + \dot{m}_{\mathrm{EP}}W_{\mathrm{EP}}\right)}{\dot{m}_a\,(H_b - H_a) + \dot{m}_c\,(H_d - H_c)}$$

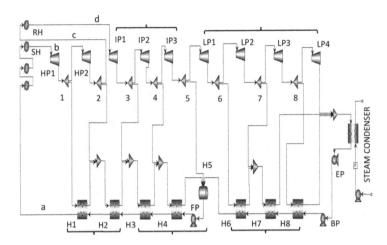

Fig. 1.29 Scheme adopted for calculating the basic thermodynamic characteristics of the cycle in Fig. 1.25

Table 1.4 Basic assumed parameters for the calculations of the steam Rankine cycle of Fig. 1.29

Maximum temperature	550 °C
Reheating temperature	550 °C
Turbine inlet pressure	170 bar
Reheating pressure	41.5 bar
Condensation pressure	0.043 bar
Efficiency of the turbines	0.77 (HP1) 0.8 (HP2) 0.85 (HP3) 0.88 (for the IP and LP cylinders)
Efficiency of the pumps	0.85
Minimum temperature difference in the closed feedwater heaters	5 °C
Subcooling degree at the output of the open feedwater heaters	2 °C
Subcooling degree on the hot side of the closed feedwater heaters	5 °C
Pressure drops: All in the steam generator (for simplicity)	60 bar

Assuming $\dot{m}_a = 1$ kg/s, we obtain a turbine power equal to 1,202.4 kW and an input from the pumps equal to 29.69 kW. The thermal power entering the cycle is 2,505.4 kW and the cycle efficiency is $\eta = 0.468$.

The ratio between the useful horsepower and the power necessary at the pumps 1,172.7/29.68 = 39.5 is very high. This is due to the modest specific work required for the compression of the liquid and the high enthalpy drop on the turbines.

The expansion ratios for the various turbine groups are as follows: 4 for the high-pressure turbines, 6.15 for the stages of intermediate pressure and 156.98 for the stages of low pressure. The total expansion ratio is equal to 3,953. These extremely high expansion ratios correspond to an equally great variation in the flow volume, mitigated in part by the steam extractions for the regeneration: 2.86 for the HP turbines, 3.98 for the IP turbines and 62.95 for the LP turbines. The global volume ratio of expansion is 878.2.

Consequently, the steam turbines are very complex machines: with numerous stages (needed to elaborate the high enthalpy drop) and with a tendency to small flows at the first stages (with possible problems of insufficient height of blades) and very high ones at the last stages (with the need to resort to multiple flows).

The flow of cooling water to the condenser is 60.8 kg/s for every kg of steam (if the water is at 20 °C, assuming a minimum temperature difference in the condenser of 5 °C and a temperature rise in the cooling water of 5 °C). If the condensation pressure passes from 0.043 bar to 0.035 bar, the flow of water necessary for the condenser increases by 2.96 times.

The ideal efficiency is $\eta_{max} = 1 - (20+273.15)/(550+273.15) = 0.644$, and the second-law efficiency is $0.468/0.644 = 0.727$. Assuming a reference temperature T_0 of 20 °C, it is possible to calculate the energy losses (or the losses of thermal availability; see Appendix B). Exergy losses in the phases of heating, evaporation, and superheating account for about 52 % of the total efficiency loss with respect to the ideal case. Adding in the exergy losses at the condenser, we reach 60 %. The remaining dissipation of thermodynamic availability is due to the regenerators, the turbomachines, and the pressure losses.

From the above, it should be clear how the great power steam cycles are characterised by very complex plant layouts: superheating followed by resuperheating, maximum pressures of 150–180 bar even reaching supercritical levels (up to 300 bar) and numerous regenerators for preheating the feedwater to the boiler. Such complexity is justified by the high power typically required by the great steam power plants. Since the early twentieth century, the unit power of the fossil steam turbine power plants has grown from just a few MW to the present 800–1,000 MW. The nuclear power plants have reached levels of 1,000 MW in even shorter time: the first commercial power plant, Calder Hall, in Sellafield, which started operating in 1956, had an electric power of 50 MW; the EPR reactor (a third-generation reactor, pressurised, cooled and moderated by light water) has a declared power of 1,750 MWe.

The sophisticated plant layout (aimed at reducing both internal and external irreversibilities of the cycle) and the rise in pressures and maximum temperatures have led to the current high efficiencies of cycle: from values of 15–20 % in the 1900s to the modern 45–50 % of fossil steam turbine power plants. The light-water nuclear power plants, due to their low operating temperatures, currently have efficiencies around 35–37 % (but, for this question, see exercise 1.2).

In the last few decades, the water steam cycle has seen a rapid and significant acceptance in combined cycles (see [22]), as the cycle for the recovery of thermal energy at the discharge of modern gas turbines.[21] The typical steam cycles for these applications have the special characteristic of several evaporation phases (two or three), in such a way as to reduce the irreversibilities (the dissipation of the thermodynamic potential) of the heat exchange during the cooling phase of the turbine discharge gases, which constitute a heat source of variable temperature.

Technological development has made steam cycles highly versatile and meant that they are widely used in all industrial sectors. Just think, for example, of the numerous plant set-ups for steam cycles that are used in the cogeneration of electricity and heat or the power plants for the thermodynamic conversion of solar energy, the CSP (Concentrated Solar Power) plants: using a parabolic-trough collector or based on a central tower or on Fresnel reflectors. As far as the power levels are concerned, the steam cycles are practically without rivals in applications using power supplies greater than a few dozen MWe up to (as we have seen) the extremes of 1,000–1,500 MW. It is rather difficult to realise efficient steam turbines for small or modest power levels.

1.7 The Closed Gas-Turbine Cycle

The first patent concerning a heat engine based on the use of a gas turbine dates back to 1791 and carries the name of an Englishman, John Barber. In 1904 Franz Stolze, a German, built a prototype by using an axial compressor and a multistage reaction turbine mounted on the same shaft. The air from the compressor was preheated, prior to the combustion, using the hot expansion gases from the turbine. The turbomachinery used, though, did not possess the performance levels needed to achieve useful work (also given that the maximum temperatures permitted by the material in use in those years were, in all probability, not very high).

The gas cycle is also known as the Brayton cycle, after George Brayton (1830–1892), an American engineer who successfully constructed a gas engine in the 1870s, operated by means of a thermodynamic cycle consisting of two isobars and two isentropes. The expansion and compression were produced by volumetric machines. Only in the twentieth century, with the availability of appropriate materials (resistant at high temperatures) and following a substantial perfecting of the fluid dynamics of the turbomachinery, was it possible to make the first working gas turbines, that is, those capable of supplying useful power.

---

[21]The Korneuburg A power station (Austria) was, in 1961, the first true combined gas-steam cycle to enter service. The power station used two gas turbines of 25 MW each and a steam turbine of 25 MW, too. The overall efficiency was around 32 %. Modern combined cycles using natural gas as fuel can reach plant efficiencies of nearly 60 % and typical power levels of 400 MW.

The first patent for a closed gas cycle is dated 1935 and refers to the prototype of a 2,000 kW pilot plant, built in Zurich in 1939: the AK-36 Plant[22] in the Escher Wyss factory.

In continuous flow machines, the work is proportional to the specific volumes.[23] To obtain useful work in a thermodynamic cycle, it is therefore necessary for the mean specific expansion volumes to be bigger than those of the compression phase. There needs to be a dilatation mechanism for the volumes, which, in the steam cycles, is represented by the vaporisation. The gas cycles do not have such an efficient instrument, in fact, if the vaporisation multiplies volumes a hundredfold, making the expansion work far greater than the compression, the gas cycles have to rely on dilatation caused by induced heating (generally in open cycles) from combustion: a threefold increase in the mean absolute temperature of the turbomachines, for example, gives an expansion work that is ideally three times that of compression and, therefore, to a certain extent, comparable to it. The differing quantities of useful work that the steam and gas cycles provide mean a radical difference in the incidence of thermodynamic losses (see Appendix C). In the case of steam, even where the dissipations in the turbine and the pump are very high, there is no risk of the expansion work approaching that of the compression, however, in the case of the gas cycle, the combined effect of the losses in diminishing the work of the turbine and increasing that of the compressor has a radical effect on the useful work, with the risk, should the machines be only moderately efficient, of nullifying it all together. For this reason, the ability to manage the high temperatures, increasing the intrinsic work difference between the turbine and the compressor, and the quality of the fluid dynamics of the turbomachines play a fundamental role in gas-turbine cycles. In modern industrial applications requiring great power, the gas-turbine cycle is, usually, in an open cycle and with internal combustion. The gas turbine is often associated with a recuperative steam cycle. The gas-turbine cycle with internal combustion (on open cycle), although attractive for the simplicity of the plant, the lightness of the turbo-compressor, the freedom of choice it gives in plant location, thanks to its ability to function without cooling water and under a wide range of climactic conditions, has a performance which is nevertheless heavily influenced by the nature of the working fluid (combustion gases) and the basic cycle pressure (atmospheric), both of which are imposed by the external environment.

The closed gas cycle has frequently been proposed for special applications: when using solid fuels or those of low quality and cost (coal, biomass, waste, syngas), waste heat recovery, solar energy, nuclear energy,[24] etc.) (see [23–25]). If, on the one hand, choosing a closed cycle means introducing certain new

---

[22]It was named after the inventors Jakob Ackeret (1898–1981), a Swiss aeronautical engineer and a pupil of A. Stodola, and Curt Keller.

[23]The specific adiabatic work is equal to the difference in enthalpy: $W = \Delta H$ (see Appendix A.2). Then, $(\partial H / \partial P)_S = 1/\rho$.

[24]In principle, both solar and nuclear sources are suitable for closed-cycle gas turbines intended for applications in space (e.g. space missions).

limitations, represented mainly by the primary heat[25] and waste heat exchangers and the need to have an appropriate cooling fluid, on the other hand, it affords two additional freedoms: (a) the choice of a working gas with preset characteristics and (b) that of a plant pressure level (the base pressure of the cycle) which affords the maximum economy for the turbogenerator-heat exchanger unit. In fact, the size of the turbomachines and the exchangers is heavily influenced by the density of the working fluid and, hence, by its pressure.

1.7.1   The Isentropic Transformation for the Perfect Gas

The ideal closed cycle is realised with ideal machines: adiabatic and isentropic compressors and turbines, heat exchangers without pressure or heat losses. The thermodynamic calculations are especially simplified if we assume that the working fluid is a perfect gas. By perfect gas, we mean a gas that meets the following conditions:

$$\frac{P}{\rho} = \frac{R}{M} T \qquad \text{volumetric equation of state}$$

$$C_P = C_V + R \quad \text{with} \quad C_V \quad \text{constant}$$

$$\gamma = \frac{C_P}{C_V}$$

with M as the molar mass of the gas and R the universal constant of the gases. The specific enthalpy and the specific entropy, therefore, are

$$H(T) - H_0 = C_P(T - T_0)$$

$$S(T, P) - S_0 = C_P \ln \frac{T}{T_0} - R \ln \frac{P}{P_0}$$

The perfect gas model approximates reasonably well to that of the ideal gas, provided the temperature variations are not extreme. For the ideal gas, the heating during isentropic compression is $(\partial T / \partial P)_S = RT / C_P$ or

$$\frac{dT}{T} = \frac{\gamma - 1}{\gamma} \frac{dP}{P}$$

---

[25]However, it should be noted that even in open gas cycles, in a combined cycle layout, the exchanger for waste heat release is, in fact, present: the recovery boiler.

Fig. 1.30 Ideal heat capacity ratio for some gases with different molecular complexity as a function of the temperature

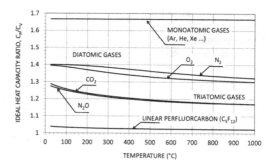

and, in the hypothesis of a perfect gas,

$$\ln \frac{T_{2'}}{T_1} = \frac{\gamma - 1}{\gamma} \ln \frac{P_2}{P_1}$$

or

$$\frac{T_{2'}}{T_1} = \left(\frac{P_2}{P_1}\right)^{\frac{\gamma-1}{\gamma}} = r^{\frac{\gamma-1}{\gamma}} \tag{1.23}$$

with T_1 and P_1 representing the temperature and pressure at the start of compression and $T_{2'}$ and P_2 the temperature and pressure at the end of isentropic compression. The ratio $P_2/P_1 = r > 1$ is the compression ratio. For an expansion

$$\frac{T_{4'}}{T_3} = \left(\frac{P_4}{P_3}\right)^{\frac{\gamma-1}{\gamma}} = \left(\frac{1}{r}\right)^{\frac{\gamma-1}{\gamma}} \tag{1.24}$$

with $T_{4'}$ representing the temperature at the end of expansion at pressure P_4, whilst expansion begins from temperature T_3 and pressure P_3. The ratio $P_3/P_4 = r > 1$ is the expansion ratio.

In general, for an ideal gas, the specific molar heat capacity C_V rises with the temperature and (with the temperature fixed) increases with the number of atoms composing the gas molecule. In fact, as the number of atoms present in the molecule rises, so does the degree of freedom available (in particular, the rotational modes and, above all, the vibrational ones). As the specific molar heat capacity C_V rises, the ratio $\gamma = (C_V + R)/C_V = 1 + R/C_V$ (with the temperature fixed) tends to decrease, reaching a limit (for high molecular complexity gas) of practically unit values (see Fig. 1.30). For a ideal gas, the ratio between the specific heats $\gamma = C_P/C_V$ depends on the temperature, but with reference to the mean values of C_V and γ, the above considerations are still valid.

Fig. 1.31 Ideal closed gas cycle with ideal gas in the thermodynamic plane T–S and diagram of the plant

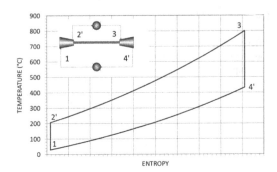

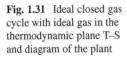

The high specific molar heat (which usually corresponds to a specific heat with near-normal values) leads directly to a peculiarity of molecular complex gases in their isentropic transformations. In fact, from the ratios (1.23) and (1.24), we deduce that the isentropic transformation, if γ approaches the unit value, also tends to become isothermal. For example, for $C_P = 200 \, \text{kJ/kmol K}$, as can be found in certain types of fluorocarbons, we have $\gamma = 1.043$ and $(\gamma - 1)/\gamma = 0.042$, which, for a pressure ratio of 5, corresponds to a temperature ratio of 1.069. For the air and helium, respectively, we get $\gamma = 1.4$ and $\gamma = 1.67$; and with a pressure ratio of 5, $T_{2'}/T_1 = 1.584$ and $T_{2'}/T_1 = 1.904$. The temperature of the fluorocarbon rises by about 7 %, that of the air by 58 % and that of the helium by 90 %.

In conclusion, if we want to have equal temperature rises, we need to impose very different compression ratios on the various different fluids.

As we shall see below, in order to get a high efficiency from air and helium gas cycles, the percentage increase of the compressor temperature (or the percentage decrease of the turbine temperature) should be approximately the same. To obtain such an increase, the compression ratio of the air cycle must be considerably higher than that of the helium cycle, with important technical consequences.

1.7.2   The Thermodynamics of the Ideal Closed Gas Cycle

The simple ideal closed cycle, with ideal gas, is shown in Fig. 1.31 in the thermodynamic plane temperature–entropy. The gas at the minimum cycle temperature T_1 is compressed from point 1 until reaching pressure P_2 at point 2'. During the adiabatic (and ideal) compression, the gas is heated up to temperature $T_{2'}$. The compression ratio $r = P_2/P_1$ is equal to the expansion ratio P_3/P_4. In the heater, the gas receives thermal energy at a constant pressure and reaches the maximum design temperature T_3. At this point, the gas expands from the maximum pressure $P_3 = P_2$ to the minimum pressure $P_4 = P_1$, cooling down to temperature $T_{4'}$. The radiator then cools the gas further to the minimum temperature T_1, closing the cycle.

Fig. 1.32 Ideal closed gas cycle with ideal gas and with a recuperative heat exchanger (the regenerator). Thermodynamic cycle in the thermodynamic plane T–S and a scheme of the plant

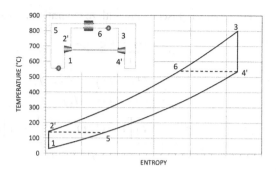

The efficiency of the simple ideal cycle with perfect gas can be calculated as

$$\eta = \frac{W_T - W_C}{Q_{in}} \tag{1.25}$$

$$= \frac{(T_3 - T_{4'}) - (T_{2'} - T_1)}{(T_3 - T_{2'})}$$

$$= \frac{(T_3 - T_{2'}) - (T_{4'} - T_1)}{(T_3 - T_{2'})}$$

$$= 1 - \frac{T_1 (T_{4'}/T_1 - 1)}{T_{2'} (T_3/T_2 - T_1)}$$

$$= 1 - \frac{T_1}{T_{2'}}$$

$$= 1 - \left(\frac{1}{r}\right)^{(\gamma-1)/\gamma}$$

since $T_{4'}/T_1 = (T_{4'}/T_1)(T_3/T_3)(T_{2'}/T_{2'}) = T_3/T_2$.
For compression ratios r lower than the value

$$r_{max}^{(\gamma-1)/\gamma} = \sqrt{\frac{T_3}{T_1}} = \sqrt{\tau} \tag{1.26}$$

the temperature $T_{4'}$ is always greater than the temperature at compression end $T_{2'}$, and it is possible to insert a recuperative heat exchanger, known as a regenerator,[26] into the cycle in order to improve the thermodynamic efficiency of the cycle, according to the diagram in Fig. 1.32.

---

[26]In reality, by "recuperative" heat exchanger, we mean an exchanger in which the two fluids (the hot one and the colder one) are physically separated by solid walls and their flow is continuous. By the term "regenerative" heat exchanger, we mean a heat exchanger made with a porous matrix, through which two fluids flow alternatively (namely, the hot one and the cold one).

For an ideal closed cycle with regeneration and a perfect gas,

$$\eta = \frac{W_T - W_C}{Q_{in}} \tag{1.27}$$

$$= \frac{(T_3 - T_{4'}) - (T_{2'} - T_1)}{(T_3 - T_6)}$$

$$= \frac{T_3\left[1 - (1/r)^{(\gamma-1)/\gamma}\right] - T_1\left[r^{(\gamma-1)/\gamma} - 1\right]}{T_3\left[1 - (1/r)^{(\gamma-1)/\gamma}\right]}$$

$$= 1 - \frac{r^{(\gamma-1)/\gamma}}{\tau}$$

since $T_6 = T_{4'}$ (see Sect. 1.7.4).

Where the compression ratio satisfies the ratio (1.26), the efficiency expressed by (1.25) is equal to that calculated by (1.27).

The useful work for the cycle (the same for the ideal cycle with regeneration and for the simple ideal cycle) is

$$\frac{W}{C_P T_1} = \tau\left[1 - \left(\frac{1}{r}\right)^{(\gamma-1)/\gamma}\right] - \left[r^{(\gamma-1)/\gamma} - 1\right] \tag{1.28}$$

Introducing the parameter $\psi = (T_{2'} - T_1)/T_1 = \left(r^{(\gamma-1)/\gamma} - 1\right)$, which represents the temperature rise in the compressor, (1.27) and (1.28) become

$$\eta = 1 - \frac{1 + \psi}{\tau} \tag{1.29}$$

$$\frac{W}{C_P T_1} = \tau\frac{\psi}{1 + \psi} - \psi \tag{1.30}$$

Here below, we shall consider just the regenerative cycle. The closed gas cycle is, in fact, always the regenerative type, largely because some heat must be released into the environment by means of a heat exchanger, and the use of this regenerator simply allows this heat release to be divided into two phases, obtaining, at the same time, a significant preheating of the working fluid on entry into the heater and, contemporarily, a decrease in the waste heat, which gives the cycle an increased thermodynamic efficiency.

The choice of the working gas for the cycle has a direct influence on the thermodynamic characteristics of the cycle and on the size of the used machinery.

Influence of the Working Fluid on Efficiency and on Useful Work in the Cycle

The two equations above (1.29) and (1.30) lead us to deduce that any working fluid (in the hypothesis, this is assumed to be a perfect gas and the machinery used ideal)

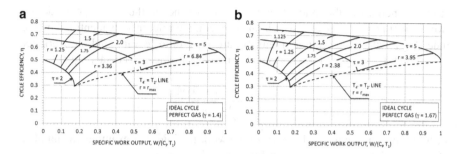

Fig. 1.33 Cycle efficiency and specific dimensionless work for ideal gas cycles with perfect gas

gives the same performance, τ and ψ being equal. What changes as the gas varies and τ remains fixed is the compression ratio necessary to obtain the desired value of ψ (see Sect. 1.7.1). The value of the specific useful work $W = W_T - W_C$ will then depend on the value T_1 and, to a large extent, on the type of gas used, through the specific heat C_P. With a fixed τ value, the cycle efficiency (in the case of an ideal cycle) diminishes with ψ, whilst the specific work output $W/(C_P T_1)$ reaches optimal where $\psi = \sqrt{\tau} - 1$, or in correspondence with the compression ratio $r = r_{max}$ of (1.26).

Figure 1.33 represents, for an ideal gas cycle with regeneration, the efficiency for the cycles and specific work $W/(C_P T_1)$ of two perfect gases (one with $\gamma = 1.4$ and the second with $\gamma = 1.67$). Fixed τ, the maximum work is obtained with the compression ratio r_{max}, when the cycle efficiency is at its minimum value. So, closed gas cycles with regeneration are generally characterised by modest compression ratios (necessary for good performances) and, consequently, by relatively modest specific work values. The effect of the molecular complexity of the working fluid is highlighted, in Fig. 1.33, by the different values of r_{max}. For example, fixed $\tau = 3$ gives $r_{max} = 6.84$ in the case of the fluid with $\gamma = 1.4$ (air) and 3.95 for the fluid with $\gamma = 1.67$ (monoatomic gas), in confirmation of what was said in Sect. 1.7.1.

Influence of the Working Fluid on the Size of the Rotating Machinery

Referring to Eqs. (1.29) and (1.30), we can see how the expansion work

$$\frac{W_T}{C_P T_1} = \tau \frac{\psi}{1 + \psi}$$

and the compression work

$$\frac{W_C}{C_P T_1} = \psi$$

are just functions of the ratios ψ and τ. With reference to the turbine, considering two distinct perfect gases (of which one of them, the reference, is identified by the index "0"), with fixed minimum temperature T_1, τ ratio, ψ ratio and the work per stage $(\Delta H_j)_S$, we get a ratio between the number of stages necessary equal to

$$\frac{N_T}{N_{T,0}} = \frac{\gamma}{\gamma - 1} \frac{\gamma_0 - 1}{\gamma_0} \frac{M_0}{M}$$

with M representing the molar mass of the working fluid. For example, the number of stages is about five times greater for helium than for air.

The work per stage (or the total enthalpy drop over the average stage, which is also equal to the difference in the static enthalpy $(\Delta H_j)_S$ if the stages are repeated and ideal) and the isentropic volume flow rate $(\dot{V}_{out,j})_S$ at the stage outlet define the socalled specific speed ω_s:

$$\omega_s = \frac{2\pi N}{60} \frac{\sqrt{(\dot{V}_{out,j})_S}}{(\Delta H_j)_S^{3/4}}$$

with N being the number of revs. The specific speed may be optimised and the dimensional analysis guarantees that (once optimised) the results for one turbine stage can be extended to all the machines with the same ω_s parameter, provided that these are working with fluids that have the same thermodynamic behaviour, that the geometric similarity is strictly respected and that the effect of the Reynolds number can be considered negligible.

The variation in the volume flow rate during expansion means that the ω_s values may vary considerably from one stage to another: smaller for the high-pressure stages and larger for low-pressure stages. Assuming that $(\Delta H_j)_S$ is the same for all stages, we calculate the ω_s variation between the first and last stage of a turbine operating with a perfect gas,

$$\frac{\omega_{s,N_T}}{\omega_{s,1}} = \sqrt{\frac{(\dot{V}_{out,N_T})_S}{(\dot{V}_{out,1})_S}}$$

$$= \sqrt{\frac{(\rho_{out,1})_S}{(\rho_{out,N_T})_S}}$$

The density for a perfect gas is calculated, once the pressure and temperature have been found from the volumetric equation of state $\rho = PM/RT$. For the final stage of the turbine, $T = T_{4'}$ and $P = P_4$ and $(\rho_{out,N_T})_S = P_4 M/RT_{4'}$. For the first stage,

$$T = T_3 - \frac{(\Delta H_j)_S}{C_P}$$

$$= T_3 \left(1 - \frac{W_T}{C_P T_1} \frac{1}{N_T} \frac{T_1}{T_3}\right)$$

$$= T_3 \left(1 - \frac{\psi}{1 + \psi} \frac{1}{N_T}\right)$$

$$P = P_3 \left(1 - \frac{\psi}{1 + \psi} \frac{1}{N_T}\right)^{\frac{\gamma}{\gamma-1}}$$

$$(\rho_{\text{out},1})_S = \frac{P_3 M}{R T_3} \left(1 - \frac{\psi}{1 + \psi} \frac{1}{N_T}\right)^{\frac{1}{\gamma-1}}$$

Then

$$\frac{\omega_{s,N_T}}{\omega_{s,1}} = \sqrt{\frac{(\rho_{\text{out},1})_S}{(\rho_{\text{out},N_T})_S}}$$

$$= \sqrt{\frac{P_3}{P_4} \frac{T_{4'}}{T_3} \left(1 - \frac{\psi}{1 + \psi} \frac{1}{N_T}\right)^{\frac{1}{\gamma-1}}}$$

$$= \sqrt{\left(\frac{T_3}{T_{4'}}\right)^{\frac{1}{\gamma-1}} \left(1 - \frac{\psi}{1 + \psi} \frac{1}{N_T}\right)^{\frac{1}{\gamma-1}}}$$

$$= \sqrt{(1 + \psi)^{\frac{1}{\gamma-1}} \left(1 - \frac{\psi}{1 + \psi} \frac{1}{N_T}\right)^{\frac{1}{\gamma-1}}}$$

If N_T (the number of stages) is sufficiently high, $\omega_{s,N_T}/\omega_{s,1} \approx \omega_{s,\infty}/\omega_{s,1} = \lim_{N_T \to +\infty} \omega_{s,N_T}/\omega_{s,1} = \sqrt{(1 + \psi)^{\frac{1}{\gamma-1}}}$. For $\psi = 0.4$ (corresponding to a compression ratio $r = 4.3$ for a perfect gas with $\gamma = 1.3$, at $r = 3.25$ for a perfect gas with $\gamma = 1.4$ and at a compression ratio $r = 2.31$ for a perfect gas with $\gamma = 1.67$), we get the following: $\omega_{s,\infty}/\omega_{s,1} = 1.75$ for $\gamma = 1.3$ (triatomic gas, carbon dioxide), 1.52 for $\gamma = 1.4$ (diatomic gas, nitrogen), and 1.28 for $\gamma = 1.67$ (monoatomic gas, helium).

The flow areas at the final-stage outlet A_{out,N_T} and at the first-stage outlet $A_{\text{out},1}$ are (for similar axial speed of the gas) in the ratio

$$\frac{A_{\text{out},N_T}}{A_{\text{out},1}} = \left(\frac{\omega_{s,N_T}}{\omega_{s,1}}\right)^2$$

and with reference to the previous example, $A_{\text{out},N_T}/A_{\text{out},1} = 3$ for $\gamma = 1.3$, 2.32 for $\gamma = 1.4$ and 1.65 for $\gamma = 1.67$. The ratio between the mean diameter and the height of the blades for the final and for the first stage varies with the square root of the flow areas, that is, they are in ratio to the specific speeds. From the examples we considered, it emerges that, in the case of monoatomic perfect gas, the turbine has more limited variations in the radial dimension than with the triatomic gas (about 30 % less, for the same turbine power).

At similar turbine power and with the maximum pressure and temperature P_3 and T_3 fixed, it is possible to calculate the ratio between the number of revs of two turbines operating with two different fluids (of which one is the reference and is identified by the index "0"), assuming the peripheral speed is the same. With reference to the first stage,

$$\frac{A_{\mathrm{out},1}}{A_{\mathrm{out},1,0}} = \frac{\gamma_0}{\gamma_0 - 1} \frac{\gamma - 1}{\gamma} \frac{\mathrm{M}}{\mathrm{M}_0} \frac{(\rho_{\mathrm{out},1,0})_S}{(\rho_{\mathrm{out},1})_S}$$

$$= \frac{\gamma_0}{\gamma_0 - 1} \frac{\gamma - 1}{\gamma} \frac{\left(1 - \frac{\psi}{1+\psi} \frac{1}{N_{T,0}}\right)^{\frac{1}{\gamma_0 - 1}}}{\left(1 - \frac{\psi}{1+\psi} \frac{1}{N_T}\right)^{\frac{1}{\gamma - 1}}}$$

$$\frac{N}{N_0} = \sqrt{\frac{A_{\mathrm{out},1,0}}{A_{\mathrm{out},1}}}$$

$$= \sqrt{\frac{\gamma}{\gamma - 1} \frac{\gamma_0 - 1}{\gamma_0} \sqrt{\frac{\left(1 - \frac{\psi}{1+\psi} \frac{1}{N_T}\right)^{\frac{1}{\gamma - 1}}}{\left(1 - \frac{\psi}{1+\psi} \frac{1}{N_{T,0}}\right)^{\frac{1}{\gamma_0 - 1}}}}}$$

If N_T and $N_{T,0}$ (the number of stages for the two turbines) are sufficiently high, $N/N_0 \approx N_\infty/N_{\infty,0} = \lim_{N_T \to +\infty} N/N_0 = \sqrt{\frac{\gamma}{\gamma-1} \frac{\gamma_0-1}{\gamma_0}}$. For $\gamma_0 = 1.4$ and $\gamma = 1.67$, we get $N_\infty/N_{\infty,0} = 0.84$. For similar useful power and at the same maximum pressure and temperature, the turbine of a closed helium cycle has a number of revolutions which is 15 % lower than the corresponding turbine with nitrogen.

So far, our considerations have concerned ideal cycles operating with perfect gas. Whilst the qualitative conclusions regarding the characteristics of the turbomachinery are valid in general, the cycle efficiency will change considerably when we introduce the real machine efficiencies. For this reason, the evaluation of the real cycle performances must take into consideration: the efficiency of the machines that they contain, which are not unitary; the pressure losses, which are always present; as well as the effective thermodynamic characteristics of the working fluid (in particular, the specific heat) which, if the gas is ideal, are a function of the temperature. The power necessary at the compressor is transmitted directly by the turbine via a shaft, which introduces a loss that is taken into account by means of the η_m mechanical efficiency (assuming it to be constant, $\eta_m = 0.95$).

A gas cycle is extremely sensitive to pressure losses on the heat exchangers (and, generally speaking, to all losses) because the ratio $W/(W_T + W_C)$ is fairly modest (in the case of an ideal cycle, $W/(W_T+W_C) = [1-(1+\psi)/\tau]/[1+(1+\psi)/\tau]$ is 0.4, for $\psi = 0.4$ and $\tau = 3.21$). Therefore, it is imperative that the losses be kept to a minimum that is technologically and economically reasonable. As far as pressure losses are concerned, the thermodynamic irreversibility that they introduce, which directly penalises the efficiency, is comparable to a throttling process, and if the gas is a perfect gas, for the purposes of calculating the cycle efficiency, it has no importance where it appears. In the following examples, for the sake of simplicity, all the pressure losses will be concentrated at the compressor outlet, and their net effect will be that of reducing the expansion ratio, which will be lower than the compression ratio ($r_T = P_3/P_4 < r_C = P_2/P_1$). The pressure losses will be introduced as a fraction of the maximum pressure (the pressure at the discharge of the compressor). In this way, $P_3 = P_2(1 - \Delta P/P_2)$ and $r_T = r_C(1 - \Delta P/P_2)$.

In Sects. 1.7.3 and 1.7.4, we shall discuss the definitions of compression, expansion, and regenerator efficiency.

1.7.3 Compressor and Turbine Efficiency

With regard to thermal turbomachinery, in order to calculate the work of compression and expansion, we usually refer to the isentropic (or adiabatic) efficiency, defined as

$$\eta_C = \frac{(\Delta H)_S}{\Delta H} \qquad \text{for a compression} \tag{1.31a}$$

$$\eta_T = \frac{\Delta H}{(\Delta H)_S} \qquad \text{for an expansion} \tag{1.31b}$$

where ΔH represents the real work and $(\Delta H)_S$ the work corresponding to an ideal adiabatic transformation (an isentropic one). The two efficiencies η_C and η_T are an index of the thermo-fluid dynamic efficiency which characterises the transformation, with the definition taking into consideration the fact that, by its very nature, the turbomachinery is poorly adapted to the continuous cooling or reheating of the working fluid.[27] Equations (1.31a) and (1.31b) are easy to apply in the case of a perfect gas because the isentropic work is easily calculated. In fact,

---

[27] In fact, there is very limited surface space available inside them for heat exchange to take place, and the fluid touches them at such high speed. For this reason, in the case of compressors, when seeking to reduce the average temperatures of the transformation, the compression is interrupted once the fluid has undergone a significant reheating; the gas is sent to a heat exchanger, where its temperature is brought back to values similar to the start; the compression is then completed at a second stage (intercooled compression). Naturally, there may be more than one intercooling stage.

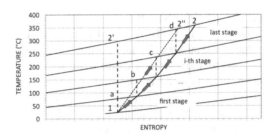

Fig. 1.34 Temperature variation during compression in the case of (i) isentropic compression from point 1 to point 2', (ii) adiabatic compression achieved with a single stage of fixed efficiency η_C from point 1 to point 2'', (iii) adiabatic compression achieved with four stages, each with an efficiency η_C from point 1 to the final point 2

$$\eta_C = \frac{T_1\left(r_C^{(\gamma-1)/\gamma} - 1\right)}{T_2 - T_1} \qquad \text{for a compression} \qquad (1.32a)$$

$$\eta_T = \frac{T_3 - T_4}{T_3\left(1 - \left(\frac{1}{r_T}\right)^{(\gamma-1)/\gamma}\right)} \qquad \text{for an expansion} \qquad (1.32b)$$

and, therefore, it is simple, given that we know η_C and η_T, to evaluate the temperatures T_2 and T_4 at the end of the transformation and the corresponding work of compression $C_P\,(T_2 - T_1)$ and of expansion $C_P\,(T_3 - T_4)$.

In reality, compression and expansion are divided up and carried out via a highly variable number of stages (in the case of traditional gas cycles, from a minimum of 2–3 compression stages and 2–3 turbine stages up to 20 compression stages and 7–10 expansion stages; with radial or axial machines, this will depend on the power and the compression/expansion ratios). In such cases, by associating each stage with an isentropic efficiency and assuming that it is the same for all the stages, we obtain—with reference to compression—that which is illustrated in Fig. 1.34.

On account of the fluid dynamic losses of the preceding stages, the typical compression stage has to work with a fluid that has a higher average temperature than it would if those previous stages had been ideal. The result is an effective compression efficiency η_{C,N_C} which, with the fixed compression ratio and the efficiency of each individual stage, becomes a function of the number of stages N_C:

$$\eta_{C,N_C} = \frac{(\Delta H)_S}{\sum_{j=1}^{N_C} \Delta H_j}$$

$$= \eta_C \frac{(\Delta H)_S}{\sum_{j=1}^{N_C} (\Delta H_j)_S}$$

Since $(\Delta H)_S < \sum_{j=1}^{N_C}\left(\Delta H_j\right)_S$, the efficiency η_{C,N_C} is lower than the efficiency η_C (a negative consequence of an unwanted "preheat" of the gas during the compression).

Where the expansion is broken up into N_T stages, there is a similar phenomenon, and we get

$$\eta_{T,N_T} = \frac{\sum_{j=1}^{N_T} \Delta H_j}{(\Delta H)_S}$$

$$= \eta_T \frac{\sum_{j=1}^{N_T} (\Delta H_j)_S}{(\Delta H)_S}$$

In such a case, the efficiency η_{T,N_T} is greater than the efficiency η_T (as the "reheating" in one stage is partially recovered in the next one).

As a result, the isentropic efficiencies η_C and η_T do not really take into account the real thermodynamic quality of the overall transformation (although, at least for a perfect gas, they are quick and easy to apply), so preference is given to the polytrophic efficiency, defined as the (local) efficiency of an infinitesimal compression or expansion (i.e. breaking the compression or expansion up into infinite stages):

$$\eta_{C,\infty} = \frac{dP/\rho}{dH} \qquad \text{for a compression} \qquad (1.33a)$$

$$\eta_{T,\infty} = \frac{dH}{dP/\rho} \qquad \text{for an expansion} \qquad (1.33b)$$

In the case of a perfect gas, the integration of (1.33a) and (1.33b) is immediate and gives

$$\frac{T_2}{T_1} = \left(\frac{P_2}{P_1}\right)^{(\gamma-1)/\eta_{C,\infty}\gamma} \qquad \text{for a compression} \qquad (1.34a)$$

$$\frac{T_4}{T_3} = \left(\frac{P_4}{P_3}\right)^{\eta_{T,\infty}(\gamma-1)/\gamma} \qquad \text{for an expansion} \qquad (1.34b)$$

In general, the specific heat of those gases comparable to ideal gases is a function of the temperature (only monoatomic gases are completely comparable to a perfect gas), whilst with a real gas, there are also effects linked to the volumetric behaviour. In such cases, integrating (1.33a) and (1.33b) is not usually simple, and the problem arises of the number of stages used in the preliminary cycle calculations in order to get expansions and compressions that have a polytrophic efficiency equal to that which was assumed. In fact, the number and type of stages to be employed are not known a priori, as this is connected to the kind of fluid used, the compression ratios, the level of power, etc.

By using similitude and dimensional analysis, it is possible to carry out a thermo-fluid dynamic cycle optimisation, using appropriate calculation codes and taking into account the effective efficiency of each stage of the compressor and

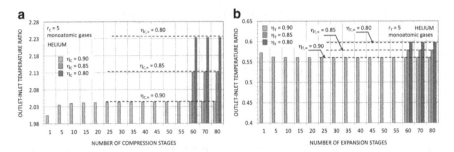

Fig. 1.35 Temperature ratio for an adiabatic compression (**a**) and expansion (**b**) as the number of stages varies and for different efficiency values. For purposes of clarity, for the cases $\eta_C = \eta_T = 0.85$ and 0.8, the figure only shows the results relative to 60, 70, and 80 stages

turbine; in a purely thermodynamic preliminary analysis, it may be enough to break up the transformation into a sufficiently high number of stages to approximate the efficiency η_{C,N_C} and η_{T,N_T} at the values $\eta_{C,\infty}$ and $\eta_{T,\infty}$.

Figure 1.35 reports the values T_2/T_1 and T_4/T_3 as the number of compression and expansion stages varies (with reference to helium, for $r_C = r_T = 5$ and for three different values of isentropic and polytropic efficiency). As the number of stages grows, the temperature ratio tends to correspond to a value corresponding to those in (1.34a) and (1.34b). When the number of stages reaches a few dozen, the difference is well below 1 %. In the cycle examples illustrated below, we assume the number of stages to be 80.

1.7.4 Efficiency of the Recuperator and the Effect of the Pressure on the Design of Heat Exchangers

The recuperative heat exchanger (see Fig. 1.32, valid for the ideal cycle) enables the gas originating from the compressor to be preheated by the sensible heat of the same gas (hotter) at the turbine outlet. The recuperative heat exchanger, aka regenerator, is characterised by a heat transfer effectiveness, defined as the ratio between the thermal power really exchanged and the thermal power that would be exchanged if the surfaces of the exchanger were infinite:

$$\epsilon_R = \frac{\dot{Q}_R}{\dot{Q}_{\max,R}} \qquad (1.35)$$

$$= \frac{\dot{m}\,(H_6 - H_2)}{\dot{m}\,(H_4 - H_{5'})}$$

$$= \frac{(H_6 - H_2)}{(H_4 - H_{5'})}$$

with H_2 enthalpy of the gas at the exit of the compressor (in the case of real compression), H_4 gas enthalpy at the turbine discharge (in the case of real expansion) and H_6 the enthalpy of the heated gas (on HP side) at the end of the regeneration. The enthalpy $H_{5'}$ is the enthalpy of the gas at the temperature T_2 but on the LP side of the regenerator.

If the gas is comparable to a perfect gas,

$$\epsilon_R = \frac{T_6 - T_2}{T_4 - T_2} \tag{1.36}$$

If $\epsilon_R = 1$, then $T_6 = T_2$. For $\epsilon_R < 1$

$$T_6 = T_2 + \epsilon_R (T_4 - T_2) \tag{1.37}$$

and the temperature difference $T_6 - T_4 = (1 - \epsilon_R)(T_2 - T_4) = T_2 - T_5$ is the minimum temperature difference reached in the exchanger, and it is constant throughout the exchanger: the two isobars P_2 and P_4, in the $T - \dot{Q}$ plane, are shown to be parallel.

It is useful to employ (1.35) (or version (1.36) for a perfect gas) in making the thermodynamic calculations for the cycle; it shows that (1.35), in the case of monophase fluids, can be expressed as a function of two nondimensional parameters and of the flow arrangement (counterflow, parallel flow, crossflow, multi-pass, periodic flow, etc.) of the heat exchanger (see [26]):

$$\epsilon_R = \epsilon (NTU_{max}, C_{min}/C_{max}, \text{flow arrangement}) \tag{1.38}$$

with C_{min} and C_{max}, respectively, as the minimum and maximum value between $\dot{m}_{cold}C_{P,cold}$ and $\dot{m}_{hot}C_{P,hot}$. The coefficient $NTU_{max} = AU/C_{min}$ is the number of transfer units (A is the heat exchanger global area and U is the overall heat transfer coefficient).

Equation (1.38) has different analytical forms according to the type of exchanger and becomes useful whenever it is necessary to select the size of the exchanger, once the overall heat transfer coefficient has been calculated.

Influence of the Working Fluid on the Size of the Recuperator

The heat recuperator is a surface heat exchanger which is normally made with compact heat exchanger surfaces. Whatever its geometry, the heat transfer and the fluid friction characteristics of the heat exchange matrix can be represented qualitatively by means of equations like (see [27])

$$StPr^{2/3} = \phi_h (Re) \tag{1.39a}$$

$$f = \phi_f (Re) \tag{1.39b}$$

where $St = h/GC_P$ is the Stanton number, $Pr = \mu C_P/k$ is the Prandtl number and $Re = 4r_h G/\mu$ is the Reynolds number[28] and $\phi_h(Re)$ and $\phi_f(Re)$ are two functions of the Reynolds number, dependent on the type of heat exchange matrix (e.g. tube-bank surfaces with plain fins, plain plait with fin surfaces)

From (1.39a) and the definition of the Stanton number St, we obtain the heat transfer coefficient h:

$$h = \frac{\mu C_P}{Pr^{2/3}} \frac{\phi_h}{4r_h} Re$$

The thermal power $\dot{Q}$ exchanged by the fluid may be calculated as

$$\dot{Q} = hA\Lambda = \frac{\mu C_P}{Pr^{2/3}} \frac{\phi_h}{4r_h} Re A \Lambda$$

where Λ is the logarithmic mean temperature difference in the heat exchanger.

The friction power $\dot{W}_f$ can be calculated as

$$\dot{W}_f = \frac{\Delta P}{\rho} GA_c = \left(\frac{f}{2} \frac{G^2}{\rho^2} \frac{A}{A_c} \right) GA_c$$

or using (1.39b) and rearranging the resulting equation,

$$\dot{W}_f = \frac{1}{2} \frac{\mu^3}{\rho^2} \left(\frac{1}{4r_h} \right)^3 Re^3 \phi_f A$$

At this point, the ratio $\dot{W}_f/\dot{Q}$, which represents the ratio between the mechanical power necessary for the circulation of the gas and the thermal power exchanged, is

$$\frac{\dot{W}_f}{\dot{Q}} = \frac{1}{2} \frac{\phi_f}{\phi_h} \left(\frac{v^2}{\Lambda} \right) \left(\frac{Pr^{2/3}}{C_P} \right) \propto \left(\frac{v^2}{\Lambda} \right) \left(\frac{1}{C_P} \right) \qquad (1.40)$$

where v is the velocity of the gas and the Prandtl number $Pr \approx 1$ for the gases.

The ratio $\dot{W}_f/\dot{Q}$ is proportional to the product of two terms: the first takes into account the working conditions (the fluid velocity and the logarithmic mean temperature difference in the heat exchanger), and the second depends on the kind of fluid used by means of its specific heat C_P. Therefore, we conclude that a low value for the ratio $\dot{W}_f/\dot{Q}$ can be obtained by limiting the speed v, adopting high logarithmic mean temperature differences and choosing a gas with a high specific heat at constant pressure.

---

[28]The parameters that define the adimensional numbers in question are $4r_h = 4A_c L/A$, $G = \dot{m}/A_c$, with r_h being the hydraulic radius, A_c the minimum total free-flow cross-sectional area, L the length of tubes and A the total surface area. G is the mass velocity. The coefficient h is the heat transfer coefficient, μ is the viscosity and k is the thermal conductivity of the gas considered.

Equation (1.40) can be further elaborated, eliminating the speed v of the gas (see [28]), considering that the thermal heat exchanged $\dot{Q}$ can also be calculated as

$$\dot{Q} = \dot{m}C_P\Delta T = (GA_c)\,C_P\Delta T$$

with ΔT as the temperature variation of the fluid. If we define the volume of the exchanger $V = A_{fr}L$, with A_{fr} being the frontal area and L the length of the heat exchanger, taking into account that the frontal area A_{fr} and the free frontal area A_c are related via the free-to-frontal area ratio $\sigma = A_c/A_{fr}$, multiplying and dividing the second term of (1.40) by $\dot{Q}$ and by V (the volume of the exchanger), we get

$$\frac{\dot{W}_f}{\dot{Q}} = \frac{1}{2}\frac{\phi_f}{\phi_h}\left(\frac{v^2}{\Lambda}\right)\left(\frac{Pr^{2/3}}{C_P}\right)\frac{\dot{Q}^2}{G^2 A_c^2 C_P^2 \Delta T^2}\left(\frac{A_c L}{\sigma}\right)^2\frac{1}{V^2}$$

or

$$\frac{\dot{W}_f}{\dot{Q}} = \frac{1}{2}\frac{\phi_f}{\phi_h}\frac{\dot{Q}^2}{\Lambda\Delta T^2}\left(\frac{L}{\sigma V}\right)^2\left(\frac{Pr^{2/3}}{\rho^2 C_P^3}\right) \tag{1.41}$$

From (1.41), it emerges that, provided the following are equal [48],

- The thermal heat exchanged (or, roughly speaking, useful power).
- Logarithmic mean temperature difference.
- Variation in fluid temperature.
- Volume occupied by the gas (and, in general terms, the cost of the exchanger, which depends almost linearly on the volume).
- The length of the exchanger to be crossed.

a good heat transfer fluid must have a low value of the ratio $\left(Pr^{2/3}/\left(\rho^2 C_P^3\right)\right)$. An efficient heat exchanger must have a low L/V ratio, that is, a large passage section A_c compared to the length of the exchanger.

For a gas $Pr \approx 1$ and for an ideal gas, the product $\rho^2 C_P^3$ depends on the temperature, the pressure and on the ratio γ of the specific heats. We then obtain

$$\frac{\dot{W}_f}{\dot{Q}} \propto \frac{\dot{Q}^2}{\Lambda\Delta T^2}\left(\frac{L}{V}\right)^2\left(\frac{T}{P}\right)^2\left(\frac{M}{R}\right)\left(\frac{\gamma-1}{\gamma}\right)^3 \tag{1.42}$$

In (1.42), the terms that depend on the operating conditions and the thermophysical properties of the gas (which intervene by means of γ and the molar mass M) have been separated. There is a clear advantage to be gained from pressurisation, since an increase in pressure leads to a significant fall in $\dot{W}_f/\dot{Q}$. An increase in the speed of the gas, on the other hand, would have the benefit of a better exchange coefficient, but at the cost of an increased $\dot{W}_f/\dot{Q}$ ratio; see (1.40).

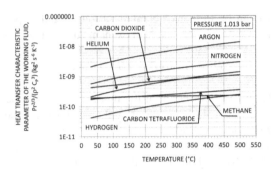

Fig. 1.36 Heat transfer parameter $Pr^{2/3}/\left(\rho^2 C_P^3\right)$ of various working fluids as a function of temperature at ambient pressure

Table 1.5 Some physical properties of the working fluids of Fig. 1.36

Fluid	Molecular weight	Critical temperature (K)	Critical pressure (bar)	Heat capacity ratio C_P/C_V (at 30 °C, 1.013 bar)
Helium, He	4.003	5.19	2.27	1.68
Argon, Ar	39.948	150.86	150.86	1.68
Xenon, Xe	131.29	289.74	58.40	1.68
Hydrogen, H_2	2.016	32.98	12.93	1.41
Nitrogen, N_2	28.014	126.2	33.98	1.41
Methane, CH_4	16.043	190.56	45.99	1.31
Carbon dioxide, CO_2	44.01	304.12	73.74	1.29
Carbon tetrafluoride, CF_4	88.005	227.51	37.45	1.16

The $\dot{W}_f/\dot{Q}$ ratio is linked to the characteristics of the fluid via its molecular complexity and its molar mass. This favours those fluids with a low γ, that is, with an elevated molecular complexity and low molar mass.

Our conclusions have a general validity and indicate a tendency, emphasising both economic and technical considerations, which are linked to the operating conditions (maximum temperatures, materials, etc.) to the cost of the working fluid, to its thermal stability and to its safety for use (nonflammability, toxicity, etc.), to its environmental friendliness: in the end, reaching the right balance between plant costs and operating costs.

Figure 1.36 shows, for certain gases at ambient pressure, the values for the parameter $\chi = Pr^{2/3}/\left(\rho^2 C_P^3\right)$ as a function of the temperature. Table 1.5 lists the molar mass, data for the critical point and the ratio $\gamma = C_P/C_V$ for specific heats at ambient temperature and ambient pressure. It is hydrogen (with the minimum molar mass) that has the best values for the χ parameter; methane (CH_4) and carbon tetrafluoride (CF_4) are very similar, carbon dioxide and helium are equivalent with temperatures over 200–250 °C and argon (a monoatomic gas with a high molar mass) is the worst of the fluids considered. As we have already observed, since χ is inversely proportional to the square of the density, an increase in the pressure will produce a more than proportional fall in its value, giving significant benefits for the

Fig. 1.37 Cycle efficiency for various working fluids as a function of the compression fractional temperature increase ψ for various maximum cycle temperatures T_3

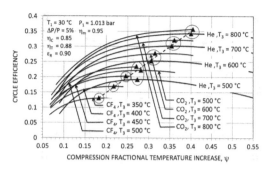

$\dot{W}_f/\dot{Q}$ ratio. In the specific case of the recuperator in a closed gas cycle, there is a great difference in the fluid density of the two branches, so the size and choice of the final geometry of the exchanger, which must obviously take into account the presence of both fluids, will generally be quite complicated.

1.7.5 The Thermodynamic of the Real Cycles

Figure 1.37 reports the efficiency of real cycles operating with various working fluids: helium, carbon dioxide, and carbon tetrafluoride. The carbon tetrafluoride (CF_4) has been taken as an example of a gas with medium to high molecular complexity and molar mass (see Table 1.5). The efficiencies η_C and η_T of compression and expansion (comparable to polytrophic efficiencies, given the large number of stages considered in the calculation; see Fig. 1.35) have been assumed to be 0.85 and 0.88, respectively. The efficiency of the recuperative heat exchanger (the regenerator) ϵ_R is 0.90. The pressure losses (5 % of the maximum pressure P_2) have been concentrated downstream from the compressor.

According to (1.29) and (1.30), the efficiency and specific useful work, having established the ratio between the temperature extremes of the cycle, are a function of just the parameter $\psi = \left(T_2' - T_1\right)/T_1$. However, these refer to the case of a perfect gas and are valid for an ideal cycle. The results in Fig. 1.37 show that, at least for the points of optimal efficiency, for the real cycles, too, there is, generally speaking, a close link between $\psi = (T_2 - T_1)/T_1$ and the value of optimal efficiency for the cycle, irrespective of the working fluid (having established the ratio between the temperature extremes of the cycle, the pressure losses and the machinery efficiency).

Figure 1.38a–c summarises, for a variety of working fluids and ratio of optimal compression, the fraction increase ψ of the temperature in the compressor, the cycle efficiency, and the specific work output $W/(C_P T_1)$. As the maximum temperature T_3 increases, so does the ψ ratio of the optimal efficiency, rising from about 0.22 for $T_3 = 400\,°C$ to 0.4 when T_3 stands at $800\,°C$. Whilst the ψ values turn out to be closely related to the changes in maximum temperature, the cycle efficiency and specific work values are less so (see, e.g. cases of $T_3 = 500\,°C$): as a consequence of the varied nature of the working fluid and the varying effect of the losses.

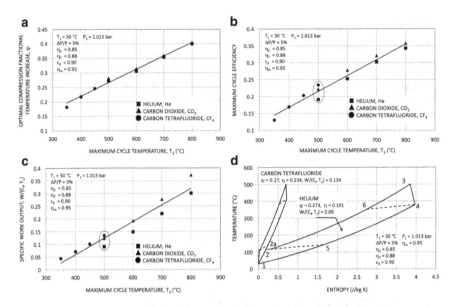

Fig. 1.38 (**a**) Compression fractional temperature increase, (**b**) cycle efficiency, and (**c**) specific work output at optimum conditions at varying maximum cycle temperature. The considered working fluids are helium, carbon dioxide, and carbon tetrafluoride. (**d**) Two cycles at the optimum ψ with $T_3 = 500\,°C$ in the T–S plane. The value of C_P assumed in the calculation of the specific work output $W/C_P T_1$ is that at $30\,°C$

For this reason, Fig. 1.38d represents two cycles, on the thermodynamic plane T–S, with the same operating conditions, but two different working fluids: helium and carbon tetrafluoride. The maximum temperature T_3 is assumed to be $500\,°C$ for both fluids (probably the limit value of thermal stability of CF_4); the minimum temperature T_1 is $30\,°C$. The minimum pressure of the cycle P_1 is the same as the atmospheric pressure. The optimal efficiency is reached in both cases when $\psi = 0.27$ (with their very different compression ratios r_C: 1.7 for helium, 5 for carbon tetrafluoride) but the efficiencies are, in these hypotheses, about 0.19 for the helium cycle and 0.23 for the CF_4 cycle.

There is an analysis of the distribution of efficiency losses with respect to the ideal value $\eta_{\max} = 1 - T_1/T_3$ (see (1.22) and Appendix C) in Fig. 1.39. As a comparison, the figure shows the values of cycle efficiency losses $\Delta\eta_j$ for a cycle with carbon dioxide, too. The thermodynamic loss introduced by the pressure losses ($\Delta\eta_2$) is that which, passing from the case of the helium cycle to the CF_4 cycle, is subject to the greatest change: from 0.043 efficiency to 0.012 (a reduction of 72 %). The increase in the loss on the regenerator ($\Delta\eta_3$) does not compensate for the dramatic reduction in loss related to the pressure losses, and the efficiency of the CF_4 cycle is around 22 % higher than the efficiency of the helium cycle. In principle, this gives an advantage at the design stage. The reason for the reduced impact of the pressure losses in a cycle that uses a gas with high molecular complexity can be shown by estimating the efficiency loss $\Delta\eta_2$ in the case of a perfect gas:

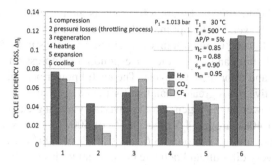

Fig. 1.39 Efficiency losses for gas cycles with optimal operating conditions, using different working fluids

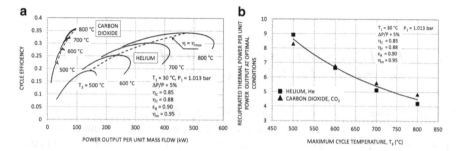

Fig. 1.40 (a) Cycle efficiency as a function of the useful power per unit flow at various maximum temperatures of the cycle. (b) Ratio between the thermal power recuperated at the regenerator and the useful power under conditions of maximum cycle efficiency, with variation in the maximum temperature

$$\Delta\eta_2 = \frac{T_1}{\dot{Q}_{in}} \Delta\dot{S}_{G,2} \approx \frac{T_1}{C_P(T_3 - T_6)}\left(-R\ln\frac{P_{2a}}{P_2}\right) = -\frac{T_1}{T_3 - T_6}\frac{\gamma - 1}{\gamma}\ln\left(1 - \frac{\Delta P}{P_2}\frac{1}{100}\right)$$

Since, generally speaking, the temperature difference $(T_3 - T_6)$ is constant (having fixed the ψ ratio and the machinery efficiency), having also established the ratio $\Delta P/P_2$, we can see how the gases with modest γ values (high molecular complexity) have $\Delta\eta_2$ values that are inferior to the corresponding values for gases with higher γ values.

The area enclosed by the cycles in Fig. 1.38d is approximately proportional to the useful work. In fact, although $W/(C_P T_1)$ for the CF_4 cycle is higher than that of the helium cycle (see Fig. 1.38c), the power per unit mass flow is about 30 kW and 140 kW, respectively, for CF_4 and helium.

Figure 1.40a reports cycle efficiencies and the useful power per unit gas mass flow for helium cycles and carbon dioxide cycles. The maximum temperature T_3

varies from 500 °C to 800 °C. Figure 1.40b reports the thermal power recuperated per unit of useful power, for the optimal values of cycle efficiency.

With T_3 equal to 800 °C, maximum cycle efficiencies are similar (0.34 and 0.36, respectively, for helium and CO_2); the corresponding compression ratios are 2 for the helium cycle and 4 for the CO_2 cycle. The useful power per unit gas mass flow for the helium is five times greater than that for the CO_2 cycle.

The thermal power on the regenerator is invariably much greater than the useful power (see Fig. 1.40b): around 4.5 times, at T_3 of 800 °C. It grows rapidly as the maximum temperature falls (around 8.6 times at T_3 of 500 °C). As a result, the regenerator is a very difficult and delicate component to design: the thermal power that it needs to operate with is very big (compared to the useful power), and its efficiency needs to be as high as possible if good thermodynamic performance is to be guaranteed (see also Appendix C.)

1.7.6   Final Considerations on the Choice of Working Fluid and Pressure Levels in the Gas Cycle

Practical considerations such as thermal stability, chemical inertia, toxicity, and nonflammability eliminate many potential candidates among the organic fluids, and in fact, there are only a few inorganic fluids that have been considered for use in the closed Brayton cycles.

In the plants built to date (see Sect. 1.7.7), air has been the most commonly used working fluid. When planning engines for space power systems, optimisation of weight and size has generally restricted the choice of working fluid to the monoatomic gases with an average molar mass: neon, argon and mixtures of helium and xenon.

In the first generation of gas nuclear reactors, carbon dioxide and helium were used just for cooling the reactor and for generating steam in the steam generators. The idea of using direct cycles with helium has been considered on several occasions.

Previous analysis showed that the choice of fluid to be used in a closed gas cycle is made difficult by the opposing needs to be met when planning the size of the turbomachinery and the heat exchangers: if in order to minimise the number of stages in a turbomachine, a gas with a low specific heat (specific to the mass) is preferred, on the other hand, in order to minimise the mechanical power needed to circulate the gas in the heat exchangers, we need a gas with a high density and specific heat. Since the density may be manipulated to a suitable value by changing the pressure in the plant, the problem can be limited to a discussion on the role of specific heat. To summarise the above, considering the influence of the molecular structure and its molar mass separately:

The structure of the molecule could be:

- Simple (high value of $\gamma = C_P/C_V$): this is always preferred, since it makes it possible to operate with low optimised compression ratios. This represents an advantage for the turbomachinery, because it permits a moderate variation in the flow volumes and average diameters along the machine shaft and, consequently, in the typical number of revs, with the benefit of high efficiency in the turbomachinery. As far as the exchangers are concerned, a low compression ratio means minimum and maximum pressure values in the cycle have the same order of magnitude: this makes it possible to maintain the right pressure in the low-pressure side, too, and, therefore, not to penalise the exchangers in charge of the heat release (regenerator and secondary exchangers; see (1.41) and (1.42)).
- Complex (low value of parameter γ): gives a high optimised compression ratio and, consequently, a tendency towards poor efficiency in the turbomachinery (both for the significant variations in the volume flows and for the potentially supersonic speeds in the turbine) and the need for great mechanical power to circulate the gas in the heat exchangers at low pressure (for the combined effect of the specific heat and the density).

Therefore, it makes sense to restrict the analysis to the simple molecule gases, which appear preferable from both points of view. We still need to examine the effect of the molar mass of the gas, which may be:

- Low (e.g. helium) which has a relatively high specific heat, desirable for heat exchangers, but, although not affecting a priori theirs efficiency, does lead to a large number of stages in the turbomachinery.
- High (e.g. argon, xenon) which is favourable for the size of the turbomachines (small number of stages) but requires the use of very large heat exchangers.

The choice between these two opposites will be of an economic nature and depend on the size of the plant: if the plant needs great power, then the impact of the cost of the exchangers on the total plant costs will be significant and, therefore, the prevalent choice will be for helium. If the plant is on a small scale, then the heat exchangers are relatively standard components and not very expensive, but the cost of the turbo-compressor group becomes fundamental: in this case, a gas with a high molar mass is favoured. There are certain plants (e.g. in space applications; see [29]) in which the economic optimisation has led to the use of a mixture of helium and xenon, mixed in such a way as to obtain a gas with a preset molar mass.

As the pressure losses are always critical in the closed gas cycles, in principle, preference is always for a working fluid with high exchange heat transfer coefficients and modest pressure losses.

Various closed-cycle plants have used air (see Sect. 1.7.7), but the presence of oxygen in high quantities leads to corrosion and limits its use to relatively low temperatures [30]. Carbon dioxide has been employed in several gas-cooled nuclear reactors, but it has a high propensity to oxidation with common steels. Argon is not desirable on account of its limited heat transfer qualities (see Sect. 1.7.4), but it is widely available (it is around 1 % by volume in air) and could be used in small power engines. Helium is inert, but it is difficult to obtain with a high degree of purity. It may contain significant quantities of hydrogen (about 5 ppm by mass), water vapour (about 50 ppm) and air (about 75 ppm). These impurities may accelerate the phenomena of oxidation and creep [31].

In a closed gas cycle, pressurisation plays an important role in the size of the cycle components. For a rough comparison in the size of the components, let's assume (1) the temperatures and the speeds of the gas are fixed at the various points of the cycle and (2) the effects of the Reynolds number are negligible, with P_0 being the low-pressure reference value and P the pressure, higher than P_0, at which the system is pressurised. This means that at equal mass flows (or equal useful power)

$$\frac{D}{D_0} = \sqrt{\frac{P_0}{P}}$$

the typical diameters of the various plant components are reduced as the pressure is increased. In the specific case of the turbomachinery, the consequence is an increase in the number of revs

$$\frac{N}{N_0} = \frac{D_0}{D} = \sqrt{\frac{P}{P_0}}$$

As a result of reducing the diameters, we can design closed gas cycles with high power using smaller machinery than that in the open gas cycles. Or, with machinery of the same size, we can obtain more useful power.

As far as the heat exchangers are concerned, in particular the recuperator, see Sect. 1.7.4, generally speaking, once the geometry of the recuperator has been fixed, the pumping power is inversely proportional to the square of the fluid density (or the pressure).

Increasing the pressure also has a positive effect on the internal efficiency of the turbomachines, and this tends to increase as the Reynolds number rises [32].

The pressurisation may also be exploited to regulate the power output. In fact, if we suppose that the machine supplies the nominal power and we wish to reduce it by 50%, since in order to reduce the power, we need only half the flow of fluid circulating in the plant, leaving the useful work unchanged (which usually corresponds to the optimised solution), it is enough to halve the maximum pressure of the cycle. In this way, we halve the density of the fluid and, consequently, the circulating mass flow. Leaving the temperature extremes of the cycle untouched, there is no significant change to the efficiency. From the point of view of the plant, in order to carry out the regulation, there needs to be a tank to which the necessary quantity of gas can be added or subtracted; furthermore, it should be emphasised that, during operation with a partial load, the heat exchangers will be super-sized and can therefore compensate for the fall in the exchange coefficient caused by the reduced pressure, thereby leaving the points of the cycle largely unchanged.

Fig. 1.41 Plants with closed
gas turbines and gas-cooled
reactors built up to 1990

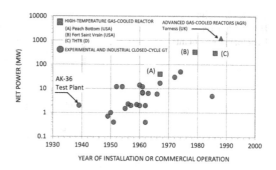

1.7.7 Examples of Several Plants that Have Been Built
and Examples of the Engines Proposed

In the industrial gas-turbine plants with closed cycles that were built between the
1950s and 1970s, the heat, deriving from combustion outside the engine, placed a
limit on the temperature of the exchangers and, consequently, on the maximum temperature of the thermodynamic cycle. The compression has always been intercooled
in order to reduce the power absorbed. Between 1939 and 1972, 23 experimental
prototypes and industrial machines were built based on a closed-cycle gas turbine,
with power supplies varying from 0.4 to 50 MW ([33] and see Fig. 1.41).

In most cases, the working fluid used was hot air. The plant at Paris in 1952, with
12 MW, was probably the most complicated turbine of this kind that was made: with
three intercoolers for the compressor, a recuperator and an intermediate expansion
heating. The machine was built with two shafts. The oil-operated combustion
chamber was turbocharged, so that smaller heating surface areas were needed. The
pipes were in austenitic steel Cr/Ni 18/8. Conditions at the input of the HP turbine
were 658 °C and 54 bar; those at the input of the LP turbine were 686 °C and
20 bar. The minimum cycle pressure was 5 bar. The high-pressure fast shaft turned
at 8,000 rpm, and the low-pressure one was synchronised at 3,000 rpm. The net
efficiency was 28.23 %.

Another important plant was built at Ravensburg, in 1956, with 2.3 MW. This
plant operated for 120,000 h without any particular problems. Its efficiency was
25 % with fuel oil and 23 % with coal. The minimum pressure was 8 bar and the
maximum 30 bar. The maximum temperature of the cycle was 660 °C.

The La Fleur engine, a helium turbine from 1962, developed by James La Fleur of
Los Angeles was probably the first helium turbine built. The engine was developed
for an air liquefaction and separation process, and the turbine, interlocked with an
inverse Brayton cycle, produced no useful electrical energy. The minimum pressure
in the system was 12.7 bar, with the maximum 18 bar, and a temperature at the start
of expansion of 650 °C. A second helium plant of 50 MW was built in 1974: the
Oberhausen II helium turbine.

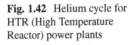

Fig. 1.42 Helium cycle for HTR (High Temperature Reactor) power plants

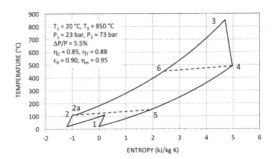

Various nuclear plants, some with high power supplies (up to 1,000 MW) were built between 1970 and 1990 (see Fig. 1.41). The gas (carbon dioxide in the advanced gas-cooled reactor (AGR) and helium in the high temperature gas-cooled reactor (HTGR) reactors) was used to cool the reactor and then to generate steam for the power-generating section by means of a heat exchanger (the steam generator).

In 1985, the Garrett Corporation (USA) made a closed gas cycle, with air as the working fluid, associated with an atmospheric fluidised bed combustor burning a low-grade fuel. The temperature at the input of the turbine was 790 °C and the plant worked apparently perfectly well, but without ever being commercialised [34].

Exercises

1.8. Figure 1.42 shows a high power helium cycle in the thermodynamic plane T–S (net power of about 1,000 MW), designed for high-temperature nuclear plants [33, p. 183]. In this example, the maximum temperature is 850 °C and the minimum has been assumed to be 20 °C. With the assumed parameters, the cycle efficiency is around 40 %.

The thermal power recuperated in the regenerator is 2.14 MW for every MW of useful power. The intercooled compression significantly reduces the compression work, increasing the useful work, to the benefit of efficiency. The total compression ratio is divided into two equal parts between the compression groups.

The geometrical dimensions of the turbine have been summarised in Fig. 1.43, case (A). There are many stages (20 axial stages) to prevent excessive peripheral speeds and to limit the centrifugal mechanical stress on the rotor blades. The number of revs is synchronous (3,000 rpm). The specific number ω_s is 0.98 for the first stage and 1.28 for the last stage. The ratio height of blade/average diameter h/D is 0.24 for the first stage and 0.29 for the last stage.

The great amount of useful power justifies the costs and the complexity of the intercooling as well as the use of helium and the demanding turbo-compression group that it entails.

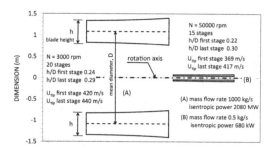

Fig. 1.43 Qualitative dimensions for turbines in helium cycles. (A) Turbine for the thermodynamic cycle of Fig. 1.42, with an isentropic power of 2,080 MW. (B) Turbine for the thermodynamic helium cycle of Fig. 1.44, isentropic power of 680 kW. h represents the blade height; D is the mean diameter

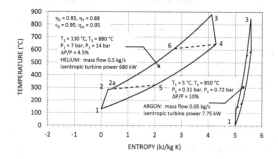

Fig. 1.44 Two thermodynamic gas cycles in closed cycles with moderate power and with different working fluids. Helium, isentropic expansion power of 680 kW. Argon, isentropic expansion power of 7.75 kW

1.9. Figure 1.44 represents, on the plane T–S, two low-powered cycles: one cycle with helium and the other with argon. The helium cycle [29], having set the power (about 200 kW), has a mass flow capacity of 0.5 kg/s but an elevated enthalpy isentropic drop (about 1,440 kJ/kg). The consequence is a small turbine (with a mean stage radius of about 6 cm; see Fig. 1.43, case (B)), with 50,000 rpm and 15 stages.

The argon cycle (useful power 3 kW [35]) is completely sub-atmospheric. The flow, which is ten times inferior to that of the helium cycle above, combined with a molar mass of the fluid around ten times greater, means that the power of this cycle is one hundredth that of the helium one. The turbine, single-stage axial with total admission, at 65,000 rpm, has an average radius of 5.8 cm and a h/D ratio (height of blade compared to average diameter of the rotor) of 0.058. The degree of reaction at the mean diameter is assumed to be 0.5.

1.10. For a closed gas cycle, assuming the parameters reported in Fig. 1.45, we find that the maximum efficiency (see the curve "cycle efficiency" of Fig. 1.45)

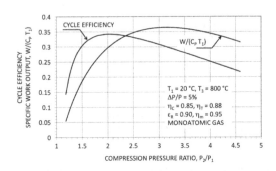

Fig. 1.45 Trends of the cycle efficiency and of the specific work output $W/(C_P T_1)$ as a function of the compression pressure ratio for monoatomic gases

Table 1.6 Several results for two closed gas cycles under conditions of maximum efficiency

	Helium	Argon
Expansion work (kJ/kg)	1,333.1	133.58
Compression work (kJ/kg)	702.73	70.388
Recuperated thermal energy (kJ/kg)	2,117.2	212.18
Recuperated energy/network	3.756	3.755
Cycle efficiency	0.359	0.359
Mass flow (kg/s)	0.0214	0.2128
Total UA of the recuperator (W/°C)	1,000.0	990.77
Total volume V of the recuperator (cm³)	1,067	10,489
Equivalent length of the side of a cube with the volume V (cm)	10.22	21.89

The useful power is 10 kW for both cases. The data and parameters that have been assumed for our calculations are to be found in Fig. 1.45

corresponds to a compression ratio $P_2/P_1 = 2$. With reference to helium and argon (two monoatomic gases), we compare and discuss the dimensions of the recuperator. We assume a useful power of 10 kW. Some results of the cycle calculation are provided in Table 1.6.

Without presuming to make detailed calculations about the regenerator, we shall limit ourselves to calculating the total volume necessary. The choice of geometry for the exchanger has been made without regard for the pressure losses. In fact, with adequate pressurisation these can be reduced at will.

Choosing the surfaces, the flow areas in the two branches, plus the need to keep pressure losses to a minimum (basically, higher on the side with the lower density fluid) and the actual form of the exchanger make a complete calculation of the regenerator fairly complicated. The complete optimisation of the regenerator (the calculation of the volume necessary for the heat transfer and the evaluation of the pressure losses) should not disregard the needs of the turbo-compressor group. At this time, though, we shall limit ourselves to a preliminary calculation of just the volume required to guarantee the exchange of the necessary thermal power.

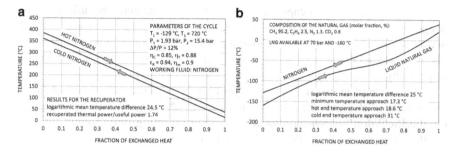

Fig. 1.46 Exercise 1.11. (**a**) Diagram of heat transfer for the recuperator in a gas cycle. (**b**) Diagram of heat transfer in the exchanger used for the regasification of LNG

The surface in question is the plain plate-fin surface #46.45T, the geometric characteristics of which, as well as the heat transfer, are reported in [26, p. 234]: the same is true for both sides of the exchanger. We assume a cross-flow configuration for the exchanger. The UA product of the global coefficient of heat transfer U and of the area A necessary for the heat transfer is known from the cycle calculations, having set the efficiency ϵ_R of the regenerator. The gas mass flow $\dot{m}$ is derived from the power-required desiderata (10 kW). Having chosen the surface type for the heat transfer, the main geometrical measurements are known, and it is possible to calculate for both flows (1) the coefficient $\alpha = A/V$, ratio of the total heat transfer area on one side to the total volume of the exchanger, and (2) the coefficient $\sigma = A_c/A_{fr}$, the ratio of the free-flow area to the frontal area. In our case, the total heat transfer areas are the same for both the hot and cold side.

Therefore, for both flows (that on the cold side and that on the hot side), having set a value for the flow-stream mass velocity G, it is possible to calculate the free-flow area $A_c = \dot{m}/G$ and then the frontal area $A_{fr} = A_c/\sigma$. The temperatures of the gas give us all the interesting physical properties of the gas, and we can evaluate the Reynolds number Re, the Stanton number St (for the surface type chosen) and the heat transfer coefficients h.

Having calculated the overall coefficient of heat transfer U (referring to the area on one of the two sides of the exchanger), from the product UA, we derive the area A (which, in our case, is the same for both sides). The volume V is found, at this point, as $V = A/\alpha$. The result is in Table 1.6: in the case of the argon cycle, the volume is ten times greater than that necessary for the helium cycle.

1.11. As a final example of a possible application of a gas turbine in a closed cycle, we consider a nitrogen cycle (see [33]) designed for the regasification of liquid natural gas (LNG, Liquefied Natural Gas). Design data for the thermodynamic cycle are reported in Fig. 1.46a, with the relative diagram of the exchanged heat in the recuperator. Assuming these parameters, the cycle efficiency is 51.8%. The waste heat, available (see Fig. 1.46b) between the temperatures T_5 and T_1 of 39.55 °C and −129 °C, heats a flow of LNG in supercritical conditions (available at 70 bar and −160 °C) up to about 20 °C. The logarithmic mean temperature difference assumed in the exchanger that regases the LNG is 25 °C, but, on account of the variation in the heat capacity of the LNG during heating, the differences in temperature between the two flows (the hot one consisting of nitrogen and the cold one consisting of LNG) vary from a minimum of 17.3 °C to a maximum of 39.2 °C. The hot and cold end temperature approach differences, result 18.6 °C and 31 °C, respectively.

A parameter of merit for a cycle designed for the regasification of natural gas is the net (electrical) power generated per unit flow of LNG, expressed in MW/(kg/s): the Specific Power Performance Parameter, $SPP = \dot{W}/\dot{m}_{LNG} = \Delta H_{NLG} \times \eta/(1-\eta)$. For this example, we have $SPP = 0.826$ MW/(kg/s).

1.8   The Stirling Engine

The Stirling engine forms part of the family of "air engines" or "hot-air engines". Inside it, a mass of gas (air, in the earliest engines; nowadays, nitrogen, helium or hydrogen at high pressure) expands and contracts in a cycle, as it is heated then cooled, producing mechanical work.

The Stirling engine incorporates a regenerator which functions as a thermodynamic sponge, alternately absorbing and releasing heat to the gas which passes through it. It is the presence of this regenerator which, theoretically, enables the Stirling engine to reach its maximum efficiency, once the two limit temperatures T_H and T_0 have been set (see Sects. 1.1 and 1.5, Fig. 1.11). The engine operates in a closed cycle.

Reverend Robert Stirling[29] invented the regenerator and the new engine that used it in 1816: this was the regenerative air engine, which would subsequently bear his name. Rev. Stirling and his brother James[30] constructed several engines over the following years. Two years after the patent, the first engine was built for pumping water from a quarry in Ayrshire (Scotland).

The air engines were used as an alternative to steam engines, towards the end of the 1800s. However, the rapid development of the internal combustion engines (Otto and Diesel) led to their decline. It was only towards the end of the 1930s that the Philips Electric Company, Eindhoven, started to study them once more and created several prototypes with good efficiency and small power supply, but without any real commercial success. For example, the engine described in [36] has a power of 180 W with an engine speed of 1,500 rpm (a diagram of the engine is shown in Fig. 1.47). General Motors, under licence from Philips, subsequently built engines with power ranging between 7–8 kW and 600 kW (for locomotives and ships). That was until 1970. Then the Ford Motor Company, again under licence from Philips, took over from General Motors, but research was called off in the early 1980s.

However, the studies into the engine were never wholly abandoned: in the 1960s, William T. Beale made a significant contribution to the development of

---

[29]Robert Stirling (1790–1878), a Minister in the Church of Scotland at Galston, Ayrshire.

[30]James Stirling (1800–1876), a Scottish engineer.

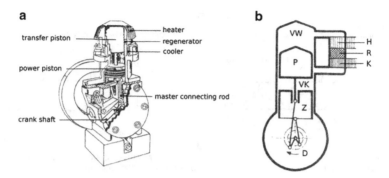

Fig. 1.47 (a) Drawing of the Philips air-engine generator MP 1002 CA. (b) Simplified diagram of the air engine. The heater (H), the regenerator (R), and the cooler (K) are shown, for clarity's sake, at one side of the cylinder and not around it (From Author [36])

the free-piston version (the free-piston Stirling engine, in which the movement of the internal parts of the engine is guaranteed by dynamic actions, without having to resort to kinematic couplings[31]); from 1964 until the 1990s, research into the Stirling engine focussed on the possibility of using it to make an artificial heart [37]; towards the end of the 1970s and up to 1990, various underwater propulsion units for non-nuclear propulsion submarines which did not require air as a comburent were designed, mainly with the contribution of the German group MAN–MWM and the Swedish company Kockums (which still has some of these engines in production) [38]. Today, there are numerous companies offering Stirling engines, above all for micro-cogeneration, often associated with the combustion (or gasification) of biomass, with electrical power of up to several dozen kW.

In fact, any Stirling engine can also operate as a refrigerator: supplying mechanical energy and removing the heat source at high temperatures. In this way, the volume designated for the expansion absorbs heat at low temperature, generating a cooling effect. By using cryogenic fluids (e.g. nitrogen, hydrogen, helium) as working fluids, we create the so-called miniature cryocoolers (with power levels of even less than 1 W). These will not be discussed here, but this is a sector in which the Stirling machine is widely used these days (although the market is relatively small).

The ideal thermodynamic cycle by which the engine operates is similar to that shown in Fig. 1.11 (the Ericsson cycle[32]) but with two isocores in place of the two isobars. An indication of how the engine works can be seen in Fig. 1.48. The basic components of the engine are as follows: a piston, which enables the exchanges of mechanical power; the "displacer", which transfers the gas in alternating fashion

---

[31] Another peculiarity of these engines is their self-starting capacity. By heating the expansion space and cooling the compression volume, the engine starts up automatically.

[32] From John Ericsson (1803–1889), an engineer born in Sweden. He emigrated to America and created numerous technically different engines.

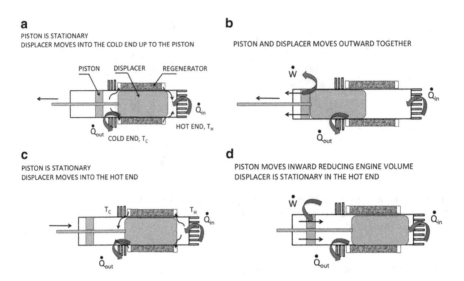

Fig. 1.48 The four phases of the Stirling engine with piston and displacer in the same cylinder ("beta" configuration). (**a**) Mass transfer stroke (from the cold to the hot side). (**b**) Expansion stroke. (**c**) Mass transfer stroke (from the hot side to the cold). (**d**) Compression stroke

from the cold to the hot head of the engine; and the regenerator, which, during the transfer of the gas from the hot volume to the cold, absorbs heat from the gas or releases heat to the gas passing through it. The presence of the regenerator is necessary if we want the engine to operate at high efficiency (although it does introduce significant pressure losses which, if they are excessive, could cancel out the useful work). The regenerator may be either external to the cylinder that contains the piston-displacer system, inside the displacer or absent. In the latter case, it is the annulus between the displacer and the cylinder that operates in place of the regenerator (by means of the cylinder and displacer walls) and it is known as regenerative annulus.

In the situation shown in Fig. 1.48a, the piston is stationary at its upper dead point, and the displacer transfers the gas from the cold head to the hot one. During the transfer (which actually happens without consuming work and, in the ideal case, with a constant total volume), the gas is forced to pass through the regenerator, which preheats it before it is entirely displaced from the cold space to the hot one.

The pressure of the gas in the hot space increases, as a consequence of the thermal power $\dot{Q}_{in}$ that is introduced. The piston-displacer system begins to move, supplying mechanical power $\dot{W}$ (Fig. 1.48b), reaching the lower dead point, corresponding to where the pressure will be at its lowest designed value.

When the piston reaches the end of its stroke (Fig. 1.48c), the displacer transfers the hot gas towards the cold space, and during this transfer, the gas, as it cools, releases part of its internal energy to the regenerator. When all the gas reaches the cold end of the engine, the piston presses it, increasing its pressure (Fig. 1.48d) and reaching once more the new upper dead point. At the same time as the compression,

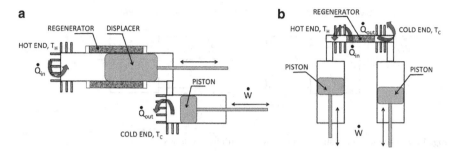

Fig. 1.49 Two further configurations of the Stirling engine. (**a**) Piston and displacer in separate cylinders ("gamma" configuration). (**b**) Two piston machine ("alpha" configuration)

there is also an extraction of thermal power $\dot{Q}_{out}$ from the engine. At this point, the thermodynamic cycle can start again. Unlike the closed cycles with external combustion based on the Rankine or Brayton cycles, the working fluid in the Stirling engines does not follow a unidirectional flow, but oscillates continuously inside the volume that constitutes the engine.

Another two classic types of engine are shown in Fig. 1.49. A great variety of mechanisms have been developed for connecting the various pistons. Several configurations of this engine, as already mentioned, have pistons that are not mechanically coupled (free pistons). Other layouts (with a hybrid coupling) have the power piston coupled mechanically with the exterior and the displacer free. A typical engine of this type is the Ringbom[33] machine [39].

The real thermodynamic cycle by which the working fluid inside the engine operates does not follow the ideal cycle at all: the pistons and the displacer do not have a discontinuous movement; the gas is not confined each time to just the well-defined volumes of expansion and compression, but distributed throughout the whole area of the engine; the expansion and the compression are not isothermal; the cooling and heating sections, which are physically distinct from the volumes of expansion and compression, introduce dead spaces[34]; the regenerator, albeit fundamental, introduces a further dead space; the regeneration is not perfect. Then there are the losses due to mechanical friction and the windage effects. The main losses inside the engine are thermal in nature, due to heat conduction from the hot expansion region to the cold compression space and dynamic (essentially the pressure losses, which depend on the number of revs and on the mean operating pressure).

The spaces in which the compression and expansion take place vary in time cyclically and simultaneously, but not in phase. They are connected by the regenerator

---

[33]Ossian Ringbom, Finnish, "subject of the Czar of Russia, residing at Borga" patented in the United States (Patent no. 856102, 4 June 1907. Application filed 17 July 1905) and in the UK (Patent no. 10675, 22 May 1906. Date of Application, 22 May 1905) his own hot-air engine.

[34]The dead spaces reduce the compression ratio and the specific useful power.

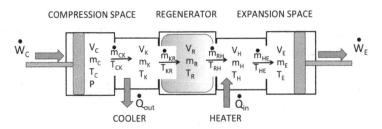

Fig. 1.50 The ideal adiabatic reference model with five volumes

and two auxiliary exchangers: the cooler and the heater. The general conceptual layout, which is valid for all the "alpha", "beta" or "gamma" configurations [40, p. 91], is shown in Fig. 1.50, with the two opposed pistons. We shall refer to it for the mono-dimensional thermodynamic analysis of the engine.

1.8.1 The Thermodynamic Analysis of the Stirling Engine

The first thermodynamic analysis of the Stirling cycle was carried out by Gustav Schmidt, a professor at the German Polytechnic Institute in Prague, in 1871 (55 years after Robert Stirling was granted his patent for the engine in 1816). Although Schmidt's analysis refers to an ideal machine, with isothermal compression and expansion, it has often been used since then as the reference for a preliminary analysis of the characteristics of any Stirling engine. Not till 1960 did Theodor Finkelstein [41], incorporate a more realistic hypothesis for adiabatic compression and expansion into the analysis, considering the spaces of expansion and compression as physically separate from those of cooling and heating. Even considering ideal exchangers (absence of pressure losses and infinite coefficients of heat exchange), Finkelstein's analysis still gives slightly more realistic results than Schmidt's analysis.[35]

Hereafter, we will refer exclusively to the adiabatic analysis; Schmidt's theory, with the perfect gas working fluid, is described and discussed in numerous textbooks, for example, in [42, 43]. The adiabatic analysis, although providing results far different from effective reality, is still useful for our purposes. It highlights the basic thermodynamic aspects linked to the choice of working fluid and the operating conditions of the cycle.

---

[35]The engine efficiency calculated by applying Schmidt's analysis (the ideal isothermal analysis), for example, coincides with that of the ideal thermodynamic cycle (the efficiency of Carnot's machine), whilst the hypothesis of adiabaticity (the ideal adiabatic analysis) leads to lower efficiency, around 20–30 % lower than that of the ideal isothermal model. However, this is still significantly higher than that of a real engine.

As mentioned (see Sect. 1.8), there are many mechanical configurations for the Stirling engine, but, in each case, the basic parameters behind the engine design are (referring to Fig. 1.50):

- The minimum temperature T_K at the cooler and maximum temperature T_H at the heater. In other words, the temperature ratio $\tau = T_H/T_K$.
- The swept compression volume $V_{SW,C}$ and the swept expansion volume $V_{SW,E}$. The corresponding swept volume ratio may be defined as $\kappa = V_{SW,C}/V_{SW,E}$.
- The total dead volume V_D, defined by means of the dead volume ratio $\zeta = V_D/V_{SW,E}$.
- The kinematic phase angle ϕ between the two pistons (see relation (1.43b)). The kinematic phase angle introduces consequently a lag between the laws of variation over time of the volumes of expansion and compression.
- The speed of the engine N.
- The pressure P: instantaneous, mean, maximum, and minimum.

In Fig. 1.50, the engine is drawn in five different volumes:

- The compression volume V_C which contains the gas at temperature T_C
- The volume of the cooler V_K
- The volume of the regenerator V_R
- The heater volume V_H
- The expansion volume V_E which contains the gas at temperature T_E

The two pistons that control the variations in the compression and expansion volumes are connected to the same shaft; the cooler, the regenerator and the heater are interposed between the two pistons. The two temperatures T_C and T_E, assuming that the two volumes V_C and V_E are adiabatic, change over time. On the other hand, the temperatures T_K and T_H are assumed to be constant and equal to the respective temperatures of the gas enclosed inside volumes V_K and V_H.

Assuming that the movement of the pistons is sinusoidal, the compression volume V_C and the expansion volume V_E vary over time, in accordance with equations [43]

$$V_C = V_{CL,C} + \frac{1}{2}V_{SW,C}\,[1.0 + \cos\omega t] \tag{1.43a}$$

$$V_E = V_{CL,E} + \frac{1}{2}V_{SW,E}\,[1.0 + \cos(\omega t + \phi)] \tag{1.43b}$$

in which $\omega = 2\pi N/60$ is the angular velocity; $V_{CL,C}$ and $V_{CL,E}$ are, respectively, the clearance spaces in the compression and expansion volumes; and $V_{SW,C}$ and $V_{SW,E}$ are the two swept volumes. In the examples below, we shall assume that (1.43a) and (1.43b) are always valid.

The total volume of the engine V_T will therefore be equal to

$$V_T = V_{CL,C} + V_{SW,C} + V_K + V_R + V_H + V_{CL,E} + V_{SW,E} \tag{1.44}$$

whilst the dead volume V_D is

$$V_D = V_{CL,C} + V_K + V_R + V_H + V_{CL,E} \tag{1.45}$$

The quantities variable over time to be calculated, in a time interval equal to the period, are:

- The pressure P in all the volumes that the engine has been drawn with
- The temperature T_C of the gas in the compression volume
- The temperature T_E of the gas in the expansion volume

Knowing P, T_E, and T_C, we can calculate the gas masses M_C, M_K, M_R, M_H, and M_E contained in the various volumes and then:

- The gas mass flow $\dot{m}_{CK}$ of the compressor to the cooler (negative if the movement is from the cooler to the compressor)
- The gas mass flow $\dot{m}_{KR}$ from the cooler to the regenerator (negative if the movement is from the regenerator to the cooler)
- The gas mass flow $\dot{m}_{RH}$ from the regenerator to the heater (negative if the movement is from the heater to the regenerator)
- The gas mass flow $\dot{m}_{HE}$ from the heater to the expansion volume (negative if the movement is in the opposite direction)

Another two parameters that need calculating are the temperature T_{RK} of the gas when it flows from the regenerator to the cooler and the temperature T_{RH} of the gas when it passes from the regenerator to the heater.

If there are no pressure losses on the heat exchangers and the regenerator, the pressure P is the same in all the volumes; if the working gas is a perfect gas and the regenerator is ideal, then $T_{RK} = T_K = T_{KR}$ and $T_{RH} = T_H = T_{HR}$.

Therefore, the useful work in the cycle is $W = | W_E | - | W_C | = \oint (| \dot{W}_E | - | \dot{W}_C |) \, dt$, and the cycle efficiency can be calculated as $\eta = W / \oint \dot{Q}_H dt = W / Q_H$.

The calculation model in the following described and used is not intended to be an instrument for making a detailed design of an engine. Its main purpose is to facilitate discussion about the thermodynamic effects and to emphasise (if only in a qualitative way) the effect of fluid dynamic losses on performance (power and efficiency) of the cycle, by means of a few global parameters.

The Ideal Engine

In the ideal case, the balance of energy applied to the compression and expansion volumes, ignoring the kinetic and gravitational energy and assuming the enthalpy, internal energy, and density function of the pressure P and the temperature, gives (see Appendix A.2 and Fig. 1.50)

$$\frac{dU_{\text{tot,C}}}{dt} = \dot{Q}_{\text{C}} + \dot{W}_{\text{C}} - \dot{m}_{\text{CK}} H_{\text{CK}} \quad \text{for the compression space}$$

$$\dot{Q}_{\text{C}} = 0.0$$

$$\dot{W}_{\text{C}} = -P \frac{dV_{\text{C}}}{dt}$$

$$U_{\text{tot,C}} = M_{\text{C}} U_{\text{C}}$$

$$M_{\text{C}} = \rho_{\text{C}} V_{\text{C}}$$

$$\frac{d}{dt} (\rho_{\text{C}} V_{\text{C}} U_{\text{C}}) + P \frac{dV_{\text{C}}}{dt} = -\dot{m}_{\text{CK}} H_{\text{CK}} \tag{1.46}$$

$$\frac{dU_{\text{tot,E}}}{dt} = \dot{Q}_{\text{E}} - \dot{W}_{\text{E}} + \dot{m}_{\text{HE}} H_{\text{HE}} \quad \text{for the expansion space}$$

$$\dot{Q}_{\text{E}} = 0.0$$

$$\dot{W}_{\text{E}} = P \frac{dV_{\text{E}}}{dt}$$

$$U_{\text{tot,E}} = M_{\text{E}} U_{\text{E}}$$

$$M_{\text{E}} = \rho_{\text{E}} V_{\text{E}}$$

$$\frac{d}{dt} (\rho_{\text{E}} V_{\text{E}} U_{\text{E}}) + P \frac{dV_{\text{E}}}{dt} = \dot{m}_{\text{HE}} H_{\text{HE}} \tag{1.47}$$

The mass conservation equation applied to the compression and expansion volumes gives (see Appendix A.1 and Fig. 1.50)

$$\frac{dM_{\text{C}}}{dt} = -\dot{m}_{\text{CK}} \quad \text{for the compression space}$$

$$\frac{d}{dt} (\rho_{\text{C}} V_{\text{C}}) = -\dot{m}_{\text{CK}}$$

$$\rho_{\text{C}} \frac{dV_{\text{C}}}{dt} + V_{\text{C}} \frac{d\rho_{\text{C}}}{dt} = -\dot{m}_{\text{CK}} \tag{1.48}$$

$$\frac{dM_{\text{E}}}{dt} = \dot{m}_{\text{HE}} \quad \text{for the expansion space}$$

$$\frac{d}{dt} (\rho_{\text{E}} V_{\text{E}}) = \dot{m}_{\text{HE}}$$

$$\rho_{\text{E}} \frac{dV_{\text{E}}}{dt} + V_{\text{E}} \frac{d\rho_{\text{E}}}{dt} = \dot{m}_{\text{HE}} \tag{1.49}$$

Combining (1.46) with (1.48), we get

$$\frac{dV_C}{dt} (\rho_C U_C + P - \rho_C H_{CK}) +$$

$$\frac{d\rho_C}{dt} V_C (U_C - H_{CK}) +$$

$$\frac{dU_C}{dt} \rho_C V_C = 0 \quad \text{for the compression space}$$

$$\text{with} \quad H_{CK} = H (P, T_{CK})$$

$$U_C = U (P, T_C)$$

$$\rho_C = \rho (P, T_C) \tag{1.50}$$

In the same way, combining (1.47) and (1.49), we get

$$\frac{dV_E}{dt} (\rho_E U_E + P - \rho_E H_{HE}) +$$

$$\frac{d\rho_E}{dt} V_E (U_E - H_{HE}) +$$

$$\frac{dU_E}{dt} \rho_E V_E = 0 \quad \text{for the expansion space}$$

$$\text{with} \quad H_{HE} = H (P, T_{HE})$$

$$U_E = U (P, T_E)$$

$$\rho_E = \rho (P, T_E) \tag{1.51}$$

Remembering that,

$$\frac{d\rho}{dt} = \left(\frac{\partial \rho}{\partial P}\right)_T \frac{dP}{dt} + \left(\frac{\partial \rho}{\partial T}\right)_P \frac{dT}{dt}$$

$$\frac{dU}{dt} = \left(\frac{\partial U}{\partial P}\right)_T \frac{dP}{dt} + \left(\frac{\partial U}{\partial T}\right)_P \frac{dT}{dt}$$

Assuming, for the sake of simplicity in the equations, the following equivalences

$$(\partial \rho_T)_C = \left(\frac{\partial \rho_C}{\partial T}\right)_P$$

$$(\partial \rho_T)_E = \left(\frac{\partial \rho_E}{\partial T}\right)_P$$

$$(\partial U_T)_C = \left(\frac{\partial U_C}{\partial T}\right)_P$$

$$(\partial U_T)_E = \left(\frac{\partial U_E}{\partial T}\right)_P$$

and

$$(\partial \rho_P)_C = \left(\frac{\partial \rho_C}{\partial P}\right)_T$$

$$(\partial \rho_P)_E = \left(\frac{\partial \rho_E}{\partial P}\right)_T$$

$$(\partial U_P)_C = \left(\frac{\partial U_C}{\partial P}\right)_T$$

$$(\partial U_P)_E = \left(\frac{\partial U_E}{\partial P}\right)_T$$

and defining

$$\alpha = \frac{dT_C}{dt}$$

$$\beta = \frac{dT_E}{dt}$$

$$\gamma = \frac{dP}{dt}$$

from (1.50), we get

$$\alpha = -\frac{\gamma V_C \left[(\partial \rho_P)_C (U_C - H_{CK}) + \rho_C (\partial U_P)_C\right] + \frac{dV_C}{dt} (\rho_C U_C + P - \rho_C H_{CK})}{V_C \left[(\partial \rho_T)_C (U_C - H_{CK}) + \rho_C (\partial U_T)_C\right]}$$

$$= \gamma A + B \tag{1.52}$$

from (1.51) we get

$$\beta = -\frac{\gamma V_E \left[(\partial \rho_P)_E (U_E - H_{HE}) + \rho_E (\partial U_P)_E\right] + \frac{dV_E}{dt} (\rho_E U_E + P - \rho_E H_{HE})}{V_E \left[(\partial \rho_T)_E (U_E - H_{HE}) + \rho_E (\partial U_T)_E\right]}$$

$$= \gamma C + D \tag{1.53}$$

We obtain the third differential equation, which defines γ, from the equation of conservation for the total mass contained within the engine (assuming there is no mass leakage)

$$M_C + M_K + M_R + M_H + M_E = M_{tot}$$

$$\frac{dM_C}{dt} + \frac{dM_K}{dt} + \frac{dM_R}{dt} + \frac{dM_H}{dt} + \frac{dM_E}{dt} = 0 \tag{1.54}$$

The mass balance equation applied to each single volume gives the respective variations in the mass held within the various volumes

$$\frac{dM_K}{dt} = \dot{m}_{CK} - \dot{m}_{KR} \quad \text{for the cooler}$$

$$\frac{d}{dt}(V_K \rho_K) = V_K \frac{d\rho_K}{dt} = \dot{m}_{CK} - \dot{m}_{KR} \tag{1.55}$$

$$\frac{dM_R}{dt} = \dot{m}_{KR} - \dot{m}_{RH} \quad \text{for the regenerator}$$

$$\frac{d}{dt}(V_R \rho_R) = V_R \frac{d\rho_R}{dt} = \dot{m}_{KR} - \dot{m}_{RH} \tag{1.56}$$

$$\frac{dM_H}{dt} = \dot{m}_{RH} - \dot{m}_{HE} \quad \text{for the heater}$$

$$\frac{d}{dt}(V_H \rho_H) = V_H \frac{d\rho_H}{dt} = \dot{m}_{RH} - \dot{m}_{HE} \tag{1.57}$$

Substituting in (1.54), (1.48), (1.49), (1.55), (1.56), and (1.57), we get (also using (1.52) and (1.53))

$$\gamma V_C \left[(\partial \rho_T)_C A + (\partial \rho_P)_C \right] + V_C (\partial \rho_T)_C B + \rho_C \frac{dV_C}{dt} +$$
$$\gamma V_K (\partial \rho_P)_K +$$
$$\gamma V_R (\partial \rho_P)_R +$$
$$\gamma V_H (\partial \rho_P)_H +$$
$$\gamma V_E \left[(\partial \rho_T)_E C + (\partial \rho_P)_E \right] + V_E (\partial \rho_T)_E D + \rho_E \frac{dV_E}{dt} = 0 \tag{1.58}$$

Or

$$\gamma = -\frac{1}{E} \left[\left(V_C (\partial \rho_T)_C B + \rho_C \frac{dV_C}{dt} \right) + \left(V_E (\partial \rho_T)_E D + \rho_E \frac{dV_E}{dt} \right) \right]$$

with

$$E = V_C \left[(\partial \rho_T)_C A + (\partial \rho_P)_C \right] +$$
$$V_K (\partial \rho_P)_K +$$
$$V_R (\partial \rho_P)_R +$$
$$V_H (\partial \rho_P)_H +$$
$$V_E \left[(\partial \rho_T)_E C + (\partial \rho_P)_E \right] \tag{1.59}$$

To calculate the enthalpy flows, it is indispensable to know the temperatures of the gas crossing the interfaces between the volumes. Let's suppose then that:

- The gas leaves the cooler at temperature T_K.
- The gas leaves the heater at temperature T_H.

and that [43]:

$$\text{if} \quad \dot{m}_{CK} > 0 \text{ then } T_{CK} = T_C$$
$$\text{if} \quad \dot{m}_{CK} < 0 \text{ then } T_{CK} = T_K$$

$$\text{if} \quad \dot{m}_{KR} > 0 \text{ then } T_{KR} = T_K$$
$$\text{if} \quad \dot{m}_{KR} < 0 \text{ then } T_{KR} = T_{RK}$$

$$\text{if} \quad \dot{m}_{RH} > 0 \text{ then } T_{RH} = T_{RH}$$
$$\text{if} \quad \dot{m}_{RH} < 0 \text{ then } T_{RH} = T_H$$

$$\text{if} \quad \dot{m}_{HE} > 0 \text{ then } T_{HE} = T_H$$
$$\text{if} \quad \dot{m}_{HE} < 0 \text{ then } T_{HE} = T_E$$

If the working fluid is an ideal gas, the specific heat at constant volume C_V and the specific heat at constant pressure C_P depend only on the temperature. Consequently, if the regenerator is ideal, the result will be $T_{RK} \rightarrow T_K$ and $T_{RH} \rightarrow T_H$.

The solution of the system of the differential equations (1.52), (1.53), and (1.59) gives the pressure P in the engine, the temperature T_C of the gas in the volume of compression and the temperature T_E of the gas in the volume of expansion. Having fixed the temperature T_{RK}, the solution is repeated for various T_{RH} ($\leq T_H$) until the heat energy Q_R exchanged globally between the gas and the matrix that constitutes the regenerator, over a complete cycle, is null. The volumetric properties of the working fluid must be calculated by an appropriate equation of state, and we must know the variation in the specific heat C_V with the temperature under ideal gas conditions (see Sect. 2.4).

The design parameters τ, κ, ζ and ϕ can be varied at will, in practically an infinite number of combinations, provided that the technological limits are respected. The thermodynamic efficiency grows with the ratio $\tau = T_H/T_K$ (see Sect. 1.1), but the temperatures T_H and T_K are restricted by metallurgical limits and by the availability of a cold heat reservoir, respectively. Frequently, the swept volume ratio $\kappa = V_{SW,C}/V_{SW,E}$ is assumed to be unitary. The optimal number of revs N is strictly linked to the working fluid and to the power desired: in the ideal case, the useful power increases linearly with the number of revs. Figure 1.51 reports

Fig. 1.51 Effect of the phase angle ϕ on the work parameter W^* for three different values of the dead volume ratio ζ. The three cases considered correspond to the three engines discussed in [43]

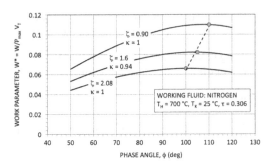

the adimensional work $W^* = W/P_{max}V_T$ based on the phase angle ϕ between the compression piston and the expansion piston, for three different values of the parameter $\zeta = V_D/V_{SW,E}$. As the ζ ratio increases, so does the dead volume, whilst the pressure variations during the cycle diminish and the useful work produced is reduced. Therefore, the smaller ζ is, the greater the useful work. The dead volume, though, should not be null, in order to guarantee the necessary exchanges of thermal energy. There is a value of the ϕ angle which maximises the work parameter W^* which, in the cases considered in the figure, is about 100°. However, the maximum of W^* as a function of ϕ is not very pronounced, and even for ϕ values around 90°, we can consider W^* as being near to its maximum value. Below, we shall always assume $\phi = 90°$.

Figure 1.52a shows, as a function of the maximum pressure, the engine efficiency and the work parameter W^* obtained by applying the adiabatic model to the case of the ideal engine. The geometric parameters assumed are those of the engine GPU-3 described in [43,44].[36] For the three working fluids considered, (nitrogen, hydrogen, and helium) neither the efficiency nor the parameter W^* varies significantly with the maximum pressure: the mean efficiency value is about 0.61 and the mean value of W^* is around 0.076. The efficiency of the corresponding ideal Carnot cycle is 0.69 (eight points higher than the value given by the ideal adiabatic analysis).

With helium (monoatomic gas) and hydrogen (diatomic gas) as the working fluids, Fig. 1.52b reports, on the P–V plane, two thermodynamic cycles with the same maximum pressure. Note (see also Fig. 1.52c) that the minimum pressure in the case of the helium engine is lower than the minimum pressure reached using hydrogen. The mean pressure of the helium engine is lower than the hydrogen one and the useful work per cycle is about 3 % lower. The helium in the (adiabatic) volume of expansion cools down significantly more than the hydrogen (see Fig. 1.52d),

---

[36]The engine GPU-3 (Ground Power Unit) is a rhombic drive Stirling engine generator, developed by General Motors in 1965 for the US Army. The values of the volumes in the engine are as follows: $V_{CL,C} = 28.68$ cm³, $V_{CL,E} = 30.52$ cm³, $V_{SW,C} = 114.13$ cm³, $V_{SW,E} = 120.82$ cm³, $V_K = 13.18$ cm³, $V_H = 70.28$ cm³ and $V_R = 50.55$ cm³. For the examples considered here, the movement of the pistons is assumed to be pure sinusoidal, in accordance with (1.43a) and (1.43b) with $\phi = 90°$.

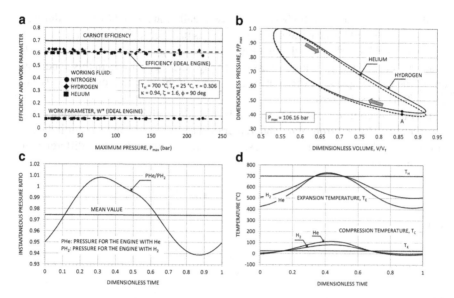

Fig. 1.52 Results of the adiabatic analysis for an ideal engine. We have assumed the geometric parameters of the engine GPU-3 (see [43,44] and the footnote 36). (**a**) Engine efficiency and work parameter for an ideal engine, with different working fluids, based on the maximum pressure. (**b**) Thermodynamic cycle on the plane P–V for two working fluids (helium and hydrogen) at the same maximum pressure. (**c**) Ratio between the instantaneous pressures for two working fluids at maximum pressure P_{max} = 106.16 bar. (**d**) Trend over time of the gas temperatures in the expansion and compression volumes for two working fluids at the same maximum pressure P_{max} = 106.16 bar

and, conversely, in the space intended for the compression (adiabatic), it tends to heat up more. It is this which gives us the greater work of compression and the lesser work of expansion in a helium cycle and, as a result, an inferior thermodynamic efficiency (about two points; see Fig. 1.52a).

The real engines give performances (useful power and efficiency) that are completely different from those of ideal engines. The complexity of the thermo-fluid dynamics, which characterises the behaviour of a Stirling engine, means that any detailed analysis of the various losses will be particularly complicated. Here below, the losses will be taken into account only through the introduction of appropriate global parameters. Without any pretensions to absolute precision, we shall consider the efficiency of the regenerator, the compression and the expansion. The pressure losses will not be counted directly as it is assumed that they will contribute anyway to increasing the work of compression and reducing that of expansion.

The Thermal Efficiency of the Regenerator

As we have seen, the regenerator is an essential component of the engine and has to be used if we wish to obtain a high efficiency. Its efficiency ϵ_R may be considered, similarly to (1.35), as the ratio between the heat really transferred to the gas during a complete cycle and the heat transferable to the gas in the case of an ideal regenerator.

Under ideal conditions, the working gas exchanges the maximum possible quantity of heat energy during its passage through the regenerator. In this condition, one of the two temperature differences at the ends of the regenerator, $(T_{RK} - T_K)$ or $(T_H - T_{RH})$, must be zero.

If, on the contrary, the regenerator is not ideal, the minimum temperature difference is different from zero, and, rather than ϵ_R, we can assume as our indicator of efficiency the parameter that represents the fractional temperature difference in the regenerator

$$\epsilon = \min(\epsilon_K, \epsilon_H) \tag{1.60}$$

with

$$\epsilon_K = \frac{T_{RK} - T_K}{T_H - T_K} \tag{1.61a}$$

and

$$\epsilon_H = \frac{T_H - T_{RH}}{T_H - T_K} \tag{1.61b}$$

As a result, in the balance equations for the energy and the mass, the temperatures of the gas contained in the regenerator and in the cooler can be assumed, respectively, to be equal to $T_R = 0.5(T_{RK} + T_{RH})$ and $T_K = 0.5(T_{RK} + T_K)$. The temperature of the gas contained in the heater is $T_H = 0.5(T_{RH} + T_H)$. In the case of a perfect gas and an ideal regenerator, $\epsilon = 0$, $T_{RK} = T_{KR} = T_K$, and $T_{RH} = T_{HR} = T_H$.

Polytropic Expansion and Compression Efficiencies

Losses caused by the viscosity of the fluid can be indirectly accounted by means of the compression and expansion efficiencies, η_E and η_C. In an ideal cycle $\eta_E = \eta_C = 1$; in a real process of compression or expansion, in accordance with the volume variation dV/dt, the work of expansion and of compression will be less and more than the ideal ones, respectively. In this way

$$\dot{W}_C = P\frac{dV_C}{dt}\frac{1}{\eta_C} \quad \text{if} \quad \frac{dV_C}{dt} < 0 \tag{1.62a}$$

$$\dot{W}_C = P\eta_C\frac{dV_C}{dt} \quad \text{if} \quad \frac{dV_C}{dt} > 0 \tag{1.62b}$$

Fig. 1.53 Power of the GPU-3 engine based on the mean operating pressure of the engine. Comparison between the values calculated and those measured and reported in [44] for hydrogen and for helium

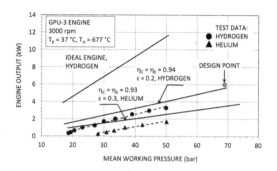

and

$$\dot{W}_{\mathrm{E}} = P\eta_{\mathrm{E}}\frac{\mathrm{d}V_{\mathrm{E}}}{\mathrm{d}t} \quad \text{if} \quad \frac{\mathrm{d}V_{\mathrm{E}}}{\mathrm{d}t} > 0 \qquad (1.63\mathrm{a})$$

$$\dot{W}_{\mathrm{E}} = P\frac{\mathrm{d}V_{\mathrm{E}}}{\mathrm{d}t}\frac{1}{\eta_{\mathrm{E}}} \quad \text{if} \quad \frac{\mathrm{d}V_{\mathrm{E}}}{\mathrm{d}t} < 0 \qquad (1.63\mathrm{b})$$

The useful work during the cycle can be calculated then in the usual way, from the work of compression and expansion.

Figure 1.53 reports the power of the engine GPU-3 as the mean operating pressure varies. The values measured (for hydrogen and helium) are those in [44]. For the calculations, which were done with an adiabatic model, temperatures T_{K} and T_{H} have been assumed as being 37 °C and 677 °C. The ideal case (see Fig. 1.53) vastly overestimates the useful power. In the case of hydrogen, assuming $\eta_{\mathrm{C}} = \eta_{\mathrm{E}} = 0.94$ and $\epsilon = 0.2$, the real power is described reasonably well: the design power, for example, at the mean pressure of about 70 bar, is 5.64 kW against 6 kW. Testing the same engine with helium gave power at about half of that with hydrogen (at the same mean operating pressure). Assuming $\eta_{\mathrm{C}} = \eta_{\mathrm{E}} = 0.93$ and $\epsilon = 0.3$, there is an acceptable agreement between the values calculated and those that were measured.

The parameters used here in order to take into account the thermodynamic losses of a real engine clearly represent just a rough and largely incomplete description of the real behaviour of an engine. Having chosen the engine and the working fluid, for example, the losses vary with the mean operating pressure and the number of revs. What is more, an engine with a preset geometry, but with different working fluids, is characterised by losses that are quantitatively different, unless operation can be guaranteed with strictly comparable fluid dynamics (as well as its geometry and kinematics). In accordance with this simple model, we need to use different values for the parameters of ϵ, η_{C} and η_{E} for the different cases we wish to consider. However, generally speaking, by making the appropriate variations to the number of revs and to the operating pressure, we should get a reasonable similarity in the fluid dynamics. Therefore, having set the reference geometry, we shall assume

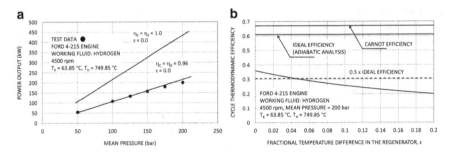

Fig. 1.54 Various results for the Ford 4-215 engine. The power values have been taken from [43]

constant values for the parameters of ϵ and $\eta_C = \eta_E$ for the various working fluids we consider. The results obtained will be purely indicative of the general thermodynamic behaviour but, for our purposes here, that will be acceptable.

Referring to Fig. 1.53, the engine operating with hydrogen at a mean pressure of 50 bar has a measured efficiency (see [44]) of around 17 %, against a calculated value of about 16 % (obtained by assuming $\eta_C = \eta_E = 0.94$ and $\epsilon = 0.2$). The efficiency values reported in [44], though, refer to the engine–combustor system and do not deduct the losses associated with the exhaust gases. In [45] the same engine, with helium, at 3,000 rpm, with a mean operating pressure of 69 bar, a maximum temperature of 650 °C and with water for cooling available at 13 °C, the energy from the gases at the combustor outlet was about 34 % of that deriving from combustion. This gave a thermodynamic efficiency of around 35 % (against an efficiency of about 12 % for the entire engine–combustor system on its own). In ref [46, pp. 73, 75] an overall brake thermal efficiency of about 27 % is reported for the same engine (operating with hydrogen, at the temperature extremes of 735/38 °C, with a mean working pressure of 69 bar).

With the purpose of assessing reasonable values for the parameters η_C, η_E and ϵ that relate to the thermodynamic performance of the engine alone, Fig. 1.54a compares several useful power values measured for the Ford 4-215 engine (from [43]), at 4,500 rpm, with the power calculated according to the adiabatic model discussed previously. Assuming $\eta_C = \eta_E = 0.96$, the useful power is clearly described, with the variation in the mean pressure of the engine. Figure 1.54b reports the thermodynamic efficiency calculated as the ϵ parameter varies for the case with a mean pressure of 200 bar. Where $\epsilon = 0.05$, the efficiency is equal to about half the efficiency of the ideal cycle. Assuming such a ratio to be reasonably representative of the quality of the engines made, from now on, we shall assume the efficiencies of compression and expansion to be 0.96 and the fractional temperature difference in the regenerator to be 0.05.

The Stirling engines, operating with a monophase gas cycle (typically, with the working fluid in the form of an ideal gas and in a similar way to the Brayton cycles), have efficiencies that diminish sharply with the maximum temperature T_H. For details, see Fig. 1.55. At a maximum temperature of 800 °C, the mean efficiency

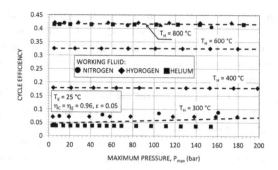

Fig. 1.55 Thermodynamic efficiency of the calculated thermodynamic cycle for different maximum temperatures T_H based on the maximum operating pressure

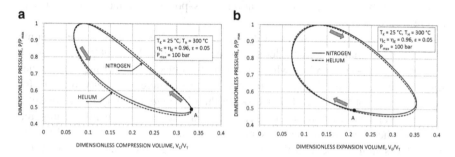

Fig. 1.56 Diagrams P–V for (**a**) the compression space and (**b**) the expansion space, for two cycles with helium and nitrogen

(taking the parameters reported in the figure for our calculations) is 58 % of that of a Carnot cycle; at 300 °C the efficiency drops to about 12 % of the ideal maximum. The ratio $W/(|W_E| + |W_C|)$ between the useful work and the sum of the two works of compression and expansion drops, for example, in the case of hydrogen with $P_{max} = 100$ bar, from 0.32 at $T_H = 800$ °C (with efficiency of 42 %) to 0.047 when $T_H = 300$ °C (efficiency of 7.2 %).

Figure 1.56a, b shows, in the P–V plane, the variations in pressure as the volume changes for the space of compression and expansion. The case we consider is that with $T_H = 300$ °C with $P_{max} = 100$ bar in Fig. 1.55. The difference in the net work of compression is the reason for the lower efficiency of the helium cycle compared to the cycle with nitrogen as its working fluid. Obviously, such comparisons and results, of a strictly thermodynamic nature, are valid purely within the scope of these extremely simplified hypotheses used for our calculations.

1.8.2 Some Final Considerations

Many of the aspects that we have drawn attention to in the closed cycles for gas turbines (Brayton or Joule cycles) are also common to Stirling cycles (consisting of two isothermal and two isocore transformations, rather than two isentropic and two isobar transformations), which are still not widely used but could have a certain growth in the future, especially in specialist sectors and for small-scale plants (generation supplies, micro-cogeneration), competing with internal combustion engines for these applications.

As we have seen, the fundamental difference from the point of view of the plant, between the gas turbine cycle and the Stirling engine, lies in the fact that the compression and expansion transformations happen in reciprocating machines (a reciprocating piston in a cylinder) rather than turbomachines; however, our considerations regarding fuels, heat exchangers, the presence of the regenerator, the choice of the working fluid, and the regulation are still essentially valid. What is more, the Stirling engine can be adapted to different fuels (natural gas and liquid fuels, but also biomass, solar energy, and waste heat from industrial processes), and it is particularly suitable for the fabrication of low-powered engines (from a few kW to a few dozen kW).

It should be noted, though, that despite the numerous attempts made by some of the greatest manufacturers, dedicating enormous development projects to the Stirling engines—researches which have never failed—the Stirling engine has still not really achieved any commercial success. There are many reasons for this situation, not the least of which being the objective difficulty in successfully competing against the reciprocating engines with internal combustion. Compared to these, the Stirling engine does present certain technological difficulties (in particular, sealing) linked to the presence of working fluids such as helium and hydrogen and to controlling the power.

Despite the objective technological difficulties, there are several engines available today, and most of the companies operating in the sector have focussed on developing engines with an electrical power of a few kW, with costs varying from 5,000 to 10,000 euro/kW.

There is a significant difference, though, in the engine layouts adopted by the producers: alpha, beta, gamma, and dual action alpha configurations, each with their own specific advantages and disadvantages. They also differ in the kind of kinematic mechanism chosen: free piston, rhombic drive, wobble joke, swash plate, rod, and crank.

As mentioned above, several producers are studying solutions that foresee the use of nonconventional fuels, such as biomass or biogas, thereby exploiting the intrinsic flexibility of the Stirling engine with regard to fuel type. Most of the engines developed work at high frequency so as to reduce the specific dimensions of the engine, partly to the detriment of the fluid dynamics and mechanical friction and to component wear, whilst the most commonly used working fluid is nitrogen, given the problems of fluid leakage associated with helium and hydrogen.

References

1. Dickinson HW (2010) A short history of the steam engine. Cambridge University Press, Cambridge
2. Savery T (1702) THE MINER'S FRIEND; or An Engine to RAISE WATER BY FIRE, DESCRIBED; and of the manner of fixing it in mines; with an account of the several other uses it is applicable unto; and an answer to the objections made against it. Printed for S. Crouch, at the Corner of Pope's Head-alley in Cornhill by W. Clowes, Stamford-street, London. Available via Google. http://books.google.com/
3. Callen HB (1985) Thermodynamics and an introduction to thermostatistics, 2nd edn. Wiley, New York
4. Müller-Steinhagen H, Trieb F (2004) Concentrating solar power. A review of the technology. Ingenia Inform Q R Acad Eng 18:43–50
5. Macchi E, Campanari S, Silva P (2005) The micro-cogeneration with natural gas. Polipress, Polytechnic of Milano, Milano (in Italian)
6. Romano M (2012) The thermoelectric plants. In: Pedrocchi E, Alimonti G (eds) Energy, environmental, and development. Progetto Leonardo, Esculapio, Bologna (in Italian)
7. Ortolani C (1984) Analysis of a modern thermoelectric plant of 300 MW. Clup, Milano (in Italian)
8. Visser WPJ, Shakariyants SA, Oostveen M (April 2011) Development of a 3 kW microturbine for CHP applications. J Eng Gas Turb Power 133:042301-1–042301-8
9. Macchi E, Pelló PM, Sacchi E (1984) Cogeneration and district heating. Thermodynamic and economical aspects. Clup, Milano (in Italian)
10. Anon (January 1985) Definitions and classification of the cogeneration processes. Draft Norm. Progetto CTI–11/146 a E02.11.146.1, Comitato Termotecnico Italiano, Milano (in Italian)
11. Zanelli F, Bonomo A, Beretta GP (2000) Fuel savings and reduction of greenhouse gases in a large waste-to-energy cogeneration facility. In: Proceedings of the 35th intersociety energy conversion engineering confference, paper AIAA-00-3059, pp 1434–1442
12. Anon (December 1999) Cogeneration total energy module - TOTEM - AdvanceD - Base 15 kW. Use and maintenance manual. TemEnergy Cogeneration Systems, Nosate (Milano)
13. Tastavi A (1990) TOTEM. Technical report. Combustion gas analysis of the cogeneration group TOTEM with and without a catalytic converter supplied with natural gas or biogas. LENI-Report-1990-001
14. Macchi E (1983) Prime movers for vapour compression heat pumps. In: Berghmns J (ed) Heat pump fundamentals. Proceedings of the NATO advanced study institute on heat pump fundamentals, Espinho, 1–12 September 1980. Martinus Nijhoff Publishers, The Hague, pp 192–225
15. Smil V (2010) Two prime movers of globalization. The history and impact of diesel engines and gas turbines. The MIT Press, Cambridge
16. Pagliano L (1989) Thermodynamic rationalization of the production of low-temperature heat. PhD Thesis, Polytechnic of Milan
17. Leyzerovich AS (May 2007) Steam turbines: how big can they get? Mod Power Syst 27(5): 43–50
18. Twardowsky A (October 2007) 858 MW supercritical extension for Belchatow. Mod Power Syst 27(10):12–17
19. Leizerovich AS (December 2009) Supercritical-pressure power plants: a progress report. Mod Power Syst 29(12):31–39
20. Kjaer S, Bugge J, Stolzenberger C (2004) Europeans still aiming for 700 °C steam. Mod Power Syst 24(11):19–25
21. Jourdain V, Herbaut P (2010) Full steam ahead for Flamanville 3 EPR turbine island construction. Mod Power Syst 30(5):17–25
22. Lozza G (2007) Gas turbine and combined cycles, 2nd edn. Progetto Leonardo, Bologna (in Italian)

23. Rohlik HE, Kofskey MG, Katsanis T (1967) Summary of NASA radial turbine research related to Brayton cycle space power systems. In: Advances in energy conversion engineering. Intersociety energy conversion engineering conference, Miami Beach, Florida, 13–17 August 1967, pp 45–54
24. Takizuka T, Takada S, Yan X, Kosugiyama S, Katanishi S, Kunitomi K (2004) R&D on the power conversion system for gas turbine high temperature reactors. Nucl Eng Des 233:329–346
25. Thomas S (2011) The pebble bed modular reactor: an obituary. Energ Pol 39:2, 431–432, 440
26. Kays WM, London AL (1998) Compact heat exchangers, 3rd edn. Krieger Publishing Company, Malabar
27. Kays WM, Crawford ME (1993) Convective heat and mass transfer, 3rd edn. McGraw-Hill, New York
28. El-Wakil MM (1962) Nuclear power engineering. McGraw-Hill, New York
29. Barret MJ (2003) Performance expectations of closed-Brayton-cycle heat exchangers in 100 kW nuclear space power systems. In: First international energy conversion engineering conference. Portsmouth, Virginia, Paper No. AIAA-2003-5956, 17–21 August 2003
30. Invernizzi CM, Iora P, Sandrini R (2011) Biomass combined cycles based on externally fired gas turbines and organic Rankine expanders. Proc IME J Power Energ 225:1066–1075
31. Lee JC, Campbell J Jr, Wright SE (1981) Closed-cycle gas turbine working fluids. J Eng Power Trans ASME 103:220–228
32. Turton RK (1995) Principles of turbomachinery, 2nd edn. Chapman & Hall, London
33. Frutschi H U (2005) Closed-cycle gas turbines. Operating experience and future potential. ASME Press, New York
34. McDonald CF (2012) Helium turbomachinery operating experience from gas turbine power plants and test facilities. Appl Therm Eng 44:108–142
35. McCormick JE, Redding TE (1967) 3 kW Recuperated closed Brayton-cycle electrical power system. In: Advances in energy conversion engineering Intersociety energy conversion engineering conference, Miami Beach, Florida, 13–17 August 1967, pp. 1–7
36. Anon (1951) Philips air-engine generator MP 1002 CA. Directions for use. Royal Philips Electronics
37. Ross B, Dudenhoefer JE (1991) Stirling machine operating experience. In: 26th intersociety energy conversion engineering conference, Boston, 4–9 August 1991
38. Hawley JG, Ashcroft SJ, Patrick MA (1994) Advanced underwater power systems. Proc IME J Power Energ 208:37–45
39. Senft JR (1993) Ringbom stirling engines. Oxford University Press, New York
40. Finkelstein T, Organ AJ (2001) Air engines. ASME Press, The American Society of Mechanical Engineers, New York
41. Finkelstein T (1960) Generalized thermodynamic analysis of Stirling engine. SAE Paper, No. 118B
42. Walker G (1980) Stirling engines. Clarendon Press/Oxford University Press, Oxford/New York
43. Urieli I, Berchowitz DM (1984) Stirling cycle engine analysis. Adam Hilger Ltd, Techno House, Redcliffe Way, Bristol
44. Cairelli JE, Thieme LG, Walter RJ (1978) Initial test results with a single-cylinder rhombic-drive Stirling engine. DOE/NASA/1040-78/1, NASA TM-78919
45. Thieme LG (1979) Low-power baseline test results for the GPU 3 Stirling engine. DOE/NASA/1040-79/6, NASA TM-79103
46. Percival WH (1974) Historical review of Stirling engine development in the United States from 1960 to 1970. NASA CR-121097
47. Casci C (1978) Elements of fluid machines - two phase fluid machines. Masson Italia Editori, Milan (in Italian)
48. Bombarda P (2010) Some notes on the heat exchangers - characteristics and performance expectations. Polytechnic of Milan (in Italian)

Chapter 2
The Thermodynamic Properties of the Working Fluids

The most common working fluids in power generation plants are water, in steam plants, and air and the gaseous combustion products in gas turbines and in internal combustion engines. Within certain limits, the air and gases produced by combustion can be treated as ideal gases. In the case of steam, when calculating the thermodynamic properties, reference needs to be made to the specific tables and diagrams. In closed gas cycles (Brayton or Stirling) the working fluid, under the usual operating conditions, is an ideal gas (helium, hydrogen etc.). Other fluids are successfully used in Rankine cycles (the so-called ORC engines, see Chap. 3) and mixtures of water and ammonia have been adopted in the so-called Kalina cycles.

Carbon dioxide and hydrocarbons are widely used as fluids in a whole variety of industrial sectors. Sodium (in liquid phase) is used as a cooling fluid in fast nuclear reactors and other liquid metals are currently proposed for future projects concerning nuclear reactors. Binary cycles with mercury and water vapour were adopted between 1930 and 1950 (see Chap. 5).

Therefore, many different working fluids are used or potentially usable in energy conversion systems and it is scarcely ever sufficient just calculating their thermodynamic properties by means of models and simplified hypotheses (as in the case of perfect or ideal gases or incompressible liquids). When detailed thermodynamic tables are not available, it is necessary to refer to equations that will give a reasonable estimate of the true thermodynamic behaviour of a working fluid, in order to make the necessary comparisons, the performance analysis and, even, a preliminary sizing of the machinery and components that will make up the thermodynamic engines.

The systems of energy conversion may also use more than one working fluid (for example, the combined cycles) and the fluid may be present in different states of aggregation.

This chapter will summarise and discuss the behaviour of fluids with regard to their possible use as working fluids in energy conversion systems.

C.M. Invernizzi, *Closed Power Cycles*, Lecture Notes in Energy 11,
DOI 10.1007/978-1-4471-5140-1_2, © Springer-Verlag London 2013

2.1　The Thermodynamic Plane of the Substances

As we know, the reasons for the deviation of a substance from the behaviour of an ideal gas are, usually, phase changes, high densities and the dissociation of the fluid. In the case of pure substances, as the pressure gradually approaches the critical pressure P_{cr}, the two phases (liquid and vapour) become increasingly indistinguishable, and at the critical pressure and critical temperature T_{cr}, the change of phase is no longer apparent. The values of P_{cr} and T_{cr} identify a particular point, called "critical point".[1]

Referring to Fig. 2.1 (relative to water, for the sake of simplicity), a state diagram distinguishes different regions, which are characterised by a different thermodynamic behaviour of the fluid [20]:

Region 1　Within the saturation curves (the saturation dome), enclosed in a temperature range that corresponds to a modest vapour pressure. If there are no changes of state, the aeriform state closely obeys the law of ideal gases. Generally, though, the variability of the vapour quality during the transformations of technical interest and the latent heat exchanged in the vaporization or condensation phases, prevents the use of transformation equations which are valid for ideal gases.

Region 2　To the right of the saturation dome, characterised by moderate pressures and widely varying temperatures. The vapour behaves like an ideal gas at specific heats that rise with the temperature. It is the only region where it is permissible to use equations valid for ideal gases.

Region 3　Within the saturation dome and at high pressure. On account of the significant "real gas effect", we cannot apply the laws of ideal gases rigorously even in the gaseous phase.

Region 4　At high pressure, in the zone of the moderately heated vapour. The high density of the vapour leads to a noticeable, although not great, deviation in the fluid behaviour from that predicted by the laws of ideal gases; generally speaking, though, it is permissible to apply the equations valid for the latter.

Region 5　To the left of the saturation dome and at low temperature. The behaviour is typical of liquids in the proper sense, characterised by negligible mechanical compressibility and a very small, though not to be overlooked, thermal dilation.

Region 6　To the left of the limit curve, for high temperatures, but subcritical. This is a liquid whose specific volume, mechanical compressibility and thermal dilation all have unusually high values. Applying the usual equations for liquids in the proper sense, in this zone, may lead to numerous mistakes (for example, the temperature rise in an ideal compression cannot usually be overlooked). In this region, we sometimes find the feed pumps for the high-pressure regenerative steam cycles.

---

[1]A critical point also exists for fluid mixtures, although, for a prefixed composition, it is not usually the point on the saturation dome at the maximum pressure and temperature (see, for example [1, p. 306]).

Fig. 2.1 Thermodynamic diagram T-S for a substance like water. Limit curves and zones with different thermodynamic behaviour

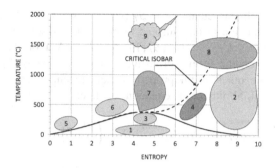

Region 7   At hypercritical pressure and temperature. Here we have a fluid with highly unusual properties, in some intermediate state between liquid and gas. Without a simple intuitive model to act as a guide for the physical compression of the phenomena, any extrapolation at that state of the liquid or gas behaviour is unadvisable.

Region 8   At high pressure and temperature. The real gas effects diminish and the steam once more conforms reasonably well to the laws for an ideal gas. In the specific case of water vapour, the high temperature (over 1,500 °C) limits the interest for this region, at least as far as industrial applications are concerned.

Region 9   At very high temperatures (2,000–3,000 °C). A new effect of the real gas appears, due to molecular dissociation, which prevents the correct use of equations valid for ideal gases. This region is of strong interest for high temperature combustion (internal combustion engines, rocket engines) where there is usually a great quantity of water vapour among the reaction products.

As a result of what we have said, only on very limited occasions and when the precision required of the calculations is not too high is it possible to use the laws of ideal gases or consider the fluid as incompressible.

Furthermore, there are essentially two different effects of a real gas. The first is apparent at high densities and in proximity to the limit curve (the saturation dome). The second, at high temperature, occurs together with the phenomena of dissociation and recombination.

2.2 The Simplified Thermodynamic Systems and the Equations of State

As we know from thermodynamics [2], the properties of a simplified thermodynamic system which is composed of r substances in a condition of stable equilibrium depend only on the internal energy U_T, the volume V_T and the number of moli $N_1, N_2 \ldots N_r$.

In particular, $S_T = S_T(U_T, V_T, N_1, N_2 \ldots N_r)$ takes the name of "fundamental equation in the entropic form", which, resolved with respect to the variable U_T, gives $U_T = U_T(S_T, V_T, N_1, N_2 \ldots N_r)$ or the "fundamental equation in the energetic form". From the expression $U_T(S_T, V_T, N_1, N_2 \ldots N_r)$ we can derive the other useful properties that are characteristic of the stable thermodynamic equilibrium or the equations of state:

$$T(S_T, V_T, N_1, N_2, \ldots N_r) = \left(\frac{\partial U_T}{\partial S_T}\right)_{V_T, N_1, N_2 \ldots N_r} \tag{2.1a}$$

$$P(S_T, V_T, N_1, N_2, \ldots N_r) = -\left(\frac{\partial U_T}{\partial V_T}\right)_{S_T, N_1, N_2 \ldots N_r} \tag{2.1b}$$

$$\mu_i(S_T, V_T, N_1, N_2, \ldots N_r) = \left(\frac{\partial U_T}{\partial N_i}\right)_{S_T, V_T, N_1, N_2 \ldots N_{r-1}} \quad \text{for } i = 1, 2 \ldots r \tag{2.1c}$$

with μ_i the chemical potential of the i-th constituent. We can also calculate the enthalpy $H_T = U_T + PV_T$, the Helmholtz free energy $A_T = U_T - TS_T$ and the Gibbs free energy $G_T = U_T - TS_T + PV_T = H_T - TS_T$.

Introducing the specific variables, referring to the mole (or mass) unit, all the extensive thermodynamic functions can be reduced to their respective specific values. For example, $U = U_T/N$, $V = V_T/N \ldots$, with $N = N_1 + N_2 + \cdots + N_r$. Defining the molar fractions $y_1 = N_1/N$, $y_2 = N_2/N \ldots$ and remembering that $y_1 + y_2 + \cdots + y_r = 1$, we get, for example, $T = T(S, V, y_1, y_2 \ldots y_r)$, $P = P(S, V, y_1, y_2 \ldots y_r)$ etc.

Since a simple thermodynamic system may have various forms of aggregation (liquid and vapour, for example), the phases are defined and a heterogeneous system is defined as a system in stable equilibrium with coexisting phases that are not separated by materialised constraints. If the phase is just one, then the system is homogeneous. As we know, a heterogeneous system has discontinuous variations in its specific properties, passing from one to another of its homogeneous parts (namely, the phases). The necessary condition for equilibrium between the various phases requires that the pressure, temperature and chemical potential of each component be the same in each phase. Under certain conditions, varying the state of equilibrium of a heterogeneous system, there may appear phase transitions (of a different order).

The limit curve in Fig. 2.1 is, for a pure fluid, the result of variations in the specific entropy S passing from the liquid to the vapour phase (example of a first order transition phase for a mono-component system[2]). The two phases—liquid and vapour—have different entropy values[3] but the variation between the value

---

[2]The phase transition is said to be of the first order if, in the transition point, the derivatives of the chemical potential show discontinuity of the first kind.

[3]The difference between specific entropies, multiplied by the temperature at which the phase transition takes place is the latent heat of evaporation.

corresponding to the liquid phase and that of the vapour phase occurs with continuity and is a function of the molar (or mass) fraction present in the various phases. Finally, we should remember that the slope of the coexistence curve for the phases on the plane $P-T$ for a mono-component heterogeneous system is represented by the Clausius–Clapeyron equation.

In the case of simple homogeneous systems, the fundamental equation $U(S, V, y_1, y_2 \ldots y_r)$ contains all the thermodynamic information. From this equation we can derive the equations of state (2.1a), (2.1b) and (2.1c). Knowing all the equations of state is the equivalent, therefore, of knowing the fundamental equation: knowing just one equation of state is to have incomplete information. However, knowing all of them bar one, it is possible to deduce the missing one, unless there is an indeterminate constant. Usually, it is relatively easy to correlate the quantities P, T and V, once we know the composition of the system, and to obtain the volumetric equation of state

$$f(P, T, V, y_1, y_2 \ldots y_r) = 0 \tag{2.2}$$

Complete knowledge of the system, though, requires a second equation which defines its thermal behaviour. This second equation of state generally correlates the specific heat at constant pressure with the temperature and pressure

$$C_P = C_P(P, T, y_1, y_2 \ldots y_r) = 0 \tag{2.3}$$

In the next section, we shall discuss the volumetric equations of state for the case of mono-component simple systems

2.3 The Volumetric Equation of State for the Pure Fluids

The numerous volumetric equations of state proposed by various authors over the years try to take into account the peculiarities of "real gases": above all, the liquid to vapour phase transitions, the volumetric behaviour in proximity to the critical point and the effects of high pressure on the density.

Of the two typical approaches to formulating the equations of state, the first based on the methods of mechanical statistics, which takes a rigorous approach to the interaction between molecules, and the second, which proposes the use of empirical or semiempirical equations based on the analysis of experimental data, we shall consider only the latter, which, for our purposes, gives simpler equations that are easy to apply.

The semiempirical and analytical equations include the so-called cubic equations. A summary of the limits and characteristics that distinguish the various equations of state is given in [3], together with suggestions and recommendations on how to develop and present any new equations of state. In [4] are presented, discussed and compared numerous cubic equations. The semiempirical equations of state generally express the pressure as the sum of two terms, a repulsive term P_R and an attractive term P_A. According to this model, no single molecule can move freely, but interacts with the molecules near to it, via the forces of cohesion and repulsion.

The general formulation of a cubic equation of state is the following:

$$P = P_R + P_A = \frac{RT}{V - b} - \frac{\Theta (V - \eta)}{(V - b)(V^2 + \delta V + \epsilon)} \tag{2.4}$$

All cubic equations derive from the equation of van der Waals,[4] the first equation of state proposed in order to describe the properties of a real gas qualitatively. Equation (2.4) is reduced to the equation of van der Waals, supposing $\delta = \epsilon = 0$, $\eta = b$ e $\Theta = a$ (constant).

The parameters which appear in (2.4) may be constants or vary with the temperature or the composition (in the case of mixtures, see Sect. 2.6). In general, they are specifics of each fluid.

Among the numerous equations proposed, one of the most used is the equation of Peng–Robinson [5]. In this, $\delta = 2b$, $\epsilon = -b^2$, $\eta = b$ and $\Theta = a_{cr}\alpha (\omega, T_r)$, with $T_r = T / T_{cr}$ (reduced temperature) and ω, a parameter called "acentric factor":

$$P = \frac{RT}{V - b} - \frac{a_{cr}\alpha (\omega, T_r)}{V (V + b) + b (V - b)} \tag{2.5a}$$

$$a_{cr} = \frac{0.457235 R^2 T_{cr}^2}{P_{cr}} \tag{2.5b}$$

$$b = \frac{0.077796 RT_{cr}}{P_{cr}} \tag{2.5c}$$

$$\alpha (\omega, T_r) = \left(1 + \kappa \left(1 - \sqrt{T_r}\right)\right)^2 \tag{2.5d}$$

$$\kappa = 0.37464 + 1.54226\omega - 0.26992\omega^2 \tag{2.5e}$$

The acentric factor, ω, was introduced by Pitzer et al. [6] as a measure of the difference in the structure of a molecule of any substance compared to that of a gas with a spherical molecule (for which ω is zero). This is defined as

$$\omega = - \log \left[P_{vp,r}\right]_{T_r=0.7} - 1 \tag{2.6}$$

---

[4]Johannes Diderik van der Waals (1837–1923). In his degree thesis, in 1873, he provided a semi-quantitative description of the phenomena of condensation and the critical point and derived the equation which bears his name. The equation of state which he developed derives from one describing the behaviour of an ideal gas, corrected in order to take into account the two special aspects of a real gas: the finite dimensions of the molecules and the intermolecular forces of attraction. In 1880, van der Waals also derived the Law of Corresponding States, showing that the equation of state which he had formulated could be expressed in a completely general form, by substituting the coefficients a and b, specifics of every fluid, with two universal parameters that are independent from the compound being considered.

Fig. 2.2 Compressibility
chart for various fluids. The
values of the compressibility
factor Z have been calculated
from the equation of state
(2.5a)

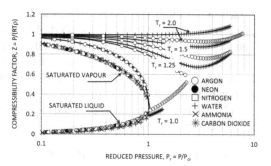

with $\left[P_{vp,r}\right]_{T_r=0.7} = \left[P_{vp}\right]_{T_r=07}/P_{cr}$ which is the reduced vapour pressure
calculated at the reduced temperature $T_r = 0.7$.

Introducing into (2.5a) the reduced variables $T_r = T/T_{cr}$, $P_r = P/P_{cr}$ and
$V_r = V/V_{cr}$, we get

$$P_r = \frac{T_r}{Z_{cr}V_r - b^\star} - \frac{a_{cr}^\star \alpha\,(\omega, T_r)}{Z_{cr}^2 V_r^2 + 2b^\star Z_{cr} - b^{\star 2}} \tag{2.7}$$

with $b^\star = 0.077796$, $a_{cr}^\star = 0.457235$ and $Z_{cr} = P_{cr}V_{cr}/RT_{cr}$ the compressibility
factor at the critical point. Equation (2.7) corresponding to the critical point ($T_r = P_r = V_r = 1$) gives $Z_{cr} = 0.307$ constant for any fluid: in the limits of
validity of (2.7), all the fluids showed the same volumetric behaviour, except for the
effects of the acentric factor ω and considering the reduced variables as variables
(see Fig. 2.2). A similar result is obtained from any cubic equation expressed by
(2.4) and goes under the name of the Law of Corresponding States. The Law
of Corresponding States, verified experimentally, may be theoretically justified
by resorting to mechanical statistics, assuming the appropriate hypotheses on the
nature of the intermolecular forces. In its simplest formulation, it is expressed
by the equation $Z = f\,(T_r, P_r)$ with two parameters, with $Z = PV/RT$ the
compressibility factor. With three parameters, the compressibility factor is more
easily estimated and, historically, as the third parameter, it is the factor Z_{cr} which
is introduced first and then, as an alternative, the acentric factor ω (see [6,7]). The
parameter Z tends towards unity when the fluid behaves in a way that tends towards
that of an ideal gas.

Starting with the Law of Corresponding States, various equations of state have
been formulated, some of them very precise and with general validity, but, they are
usually very complex. By way of example, we cite those discussed in [8,9] and the
one presented in [10, Sect. 4].

Observing that in the description of the volumetric properties of a fluid (and
mixtures of fluids) it is definitely advantageous to use models with a solid physical
base and, for this very reason, the equations of state deriving from mechanical
statistics have a great potential, many studies into the research for new formulations
of equations of state go back to appropriate molecular models, taking into account

interaction potentials that are both inter- and intra-molecular. One example of this approach is the equation PC-SAFT (Perturbed-Chain Statistical Associating Fluid Theory) with three (or more, depending on the system considered) coefficients [11].

In the considerations that follow and for the calculations in the rest of the book, for the sake of simplicity and practicality, we shall refer to the Peng–Robinson equation of state (2.5).

In (2.5a) it is possible to introduce the factor of compressibility $Z = RT/PV$ and the reduced variables T_r and P_r, obtaining

$$Z^3 - (1 - B) Z^2 + (A - 2B - 3B^2) Z - (AB - B^2 - B^3) = 0 \qquad (2.8)$$

with

$$A = a_{cr}^\star \alpha (\omega, T_r) \frac{P_r}{T_r^2} \qquad (2.9a)$$

$$B = b^\star \frac{P_r}{T_r} \qquad (2.9b)$$

Equations (2.8) with (2.9a) and (2.9b) is another form of (2.7) and allows the factor of compressibility to be calculated directly from the reduced variables T_r and P_r.

For example, if we consider nitrogen and carbon dioxide at $T = 25\,°C$ and $P = 100$ bar, the compressibility factor Z is equal to 0.989 and 0.249, respectively: nitrogen is, to all effects, an ideal gas, whilst carbon dioxide, under fixed conditions, has a density that is four times greater than that which would result from calculation with hypotheses of an ideal gas.

The Influence of Compressibility on the Work of Expansion and Compression

The volumetric effects of a real gas (represented and described by means of the compressibility factor Z), in general, reduce the molar work of compression and expansion. In fact, referring to an isentropic expansion (ideal and adiabatic) in a turbomachine from pressure P_1 to final pressure P_2:

$$W_T = \int_{P_1}^{P_2} \left(\frac{\partial H}{\partial P} \right)_S dP = \int_{P_1}^{P_2} \frac{1}{\rho} dP = \int_{P_1}^{P_2} V dP = \int_{P_1}^{P_2} Z \frac{RT}{P} dP \qquad (2.10)$$

and having set T and P, being mostly $Z < 1$, the specific work is less than that which the fluid would provide if it were an ideal gas.

Furthermore, the work of compression and expansion diminishes with the molar mass of the working fluid. Still referring to an isentropic expansion in a turbomachine:

$$W_T = \int_{P_1}^{P_2} V\,\mathrm{d}P = \int_{P_1}^{P_2} Z\frac{RT}{P}\mathrm{d}P$$

$$= \int_{P_{r,1}}^{P_{r,2}} Z\frac{T_r}{P_r} RT_{cr}\mathrm{d}P_r \quad \text{or}$$

$$\frac{W_T}{RT_{cr}} = \int_{P_{r,1}}^{P_{r,2}} Z\frac{T_r}{P_r}\mathrm{d}P_r \tag{2.11}$$

Having set the law of variation T_r as a function of P_r, the ratio W_T/RT_{cr} is constant for any fluid (generally speaking, by virtue of the Law of Corresponding States). So, as a result, the specific work is inversely proportional to the molar mass of the working fluid. The variation in the work with the molar mass of the fluid has, for example, a direct effect on the performance and size of the turbomachinery in a thermodynamic cycle with a perfect gas (see Sect. 1.7).

2.4   The Evaluation of the Thermodynamic Properties of a Real Gas

From the volumetric equation of state $f(P, V, T) = 0$ and from the equation $C_P^0 = C_P^0(T)$ for the ideal gas, it is possible to derive the thermodynamic properties that interest us. We report the necessary equations, with reference to a monophase and mono-component fluid system, following the procedure described in detail in [12, Chap. 8].

We shall start by considering the isothermal variations of a generic thermodynamic property; subsequently, we shall take into account the effects of variations in temperature.

To calculate the variations in a thermodynamic property at constant temperature, we use "departure functions", defined as functions that represent the difference between a generic thermodynamic property in the state of a real gas (at fixed temperature T and corresponding to a defined specific volume V or a pressure P) and the corresponding property for an ideal gas, at the same temperature T and pressure P but with a volume V^0 calculated by means of $V^0 = RT/P$. The choice of the two independent variables T and P or T and V is arbitrary and dictated by convenience.

Thus, if X represents any thermodynamic property (enthalpy, entropy etc.), the departure function will be defined as

$$X(T, P) - [X(T, P)]_{\text{idealgas}} = X(T, P) - X^0(T, P) \tag{2.12a}$$

$$X(T, V) - [X(T, V^0)]_{\text{idealgas}} = X(T, V) - X^0(T, V^0) \tag{2.12b}$$

Since (2.5a) is explicit in P, it is more convenient to use temperature and volume as independent variables.

Proceeding and using (2.12b), we get the departure function relative to the Helmholtz free energy:

$$A\left(T, V\right) - A^0\left(T, V^0\right) = -\int_\infty^V \left(P - \frac{RT}{V}\right) dV + RT \ln \frac{V^0}{V} \qquad (2.13)$$

Note that for A, the departure function is calculated for the entropy S as

$$\left(\frac{\partial A}{\partial T}\right)_V = -S$$

We obtain

$$S\left(T, V\right) - S^0\left(T, V^0\right) = \left[\frac{\partial}{\partial T} \int_\infty^V \left(P - \frac{RT}{V}\right) dV\right]_V + R \ln \frac{V^0}{V} \qquad (2.14)$$

From (2.13) and (2.14), we can calculate

$$U\left(T, V\right) - U^0\left(T, V^0\right) = \left(A - A^0\right) + T\left(S - S^0\right) \qquad (2.15a)$$

$$H\left(T, V\right) - H^0\left(T, V^0\right) = \left(U - U^0\right) + PV - RT \qquad (2.15b)$$

At this point, to determine the difference between the generic thermodynamic property X between two states 1 and 2 which also differ in temperature, it is possible to use

$$X\left(T_2, V_2\right) - X\left(T_1, V_1\right)) = \left[X\left(T_2, V_2\right) - X^0\left(T_2, V_2^0\right)\right]$$
$$- \left[X\left(T_1, V_1\right) - X^0\left(T_1, V_1^0\right)\right]$$
$$+ \left[X^0\left(T_2, V_2^0\right) - X^0\left(T_1, V_1^0\right)\right] \qquad (2.16)$$

in which, the first two terms used on the right side represent the departure functions at temperatures T_1 and T_2; the third term represents the difference in X under the conditions of an ideal gas, easily calculated from the specific heats of the ideal gas. For example,

$$U^0\left(T_2, V_2^0\right) - U^0\left(T_1, V_1^0\right)) = \int_{T_1}^{T_2} C_V^0 dT \qquad (2.17a)$$

$$H^0\left(T_2, V_2^0\right) - H^0\left(T_1, V_1^0\right)) = \int_{T_1}^{T_2} C_P^0 dT \qquad (2.17b)$$

$$S^0\left(T_2, V_2^0\right) - S^0\left(T_1, V_1^0\right)) = \int_{T_1}^{T_2} \frac{C_V^0}{T} dT + R \ln \frac{V_2^0}{V_1^0} \qquad (2.17c)$$

The departure functions (2.13), (2.14), (2.15a) and (2.15b) depend only on the volumetric behaviour of the fluid. The specific heat (at constant pressure and volume) comes into play only in the definition of the properties of the ideal gas.

Equation (2.13) may be expressed also as a function of the reduced variables T_r, P_r and V_r and of the compressibility factor. With reference to (2.7),

$$
\begin{aligned}
\frac{A(T, V) - A^0(T, V^0)}{RT_{cr}} &= -Z_{cr} \int_{\infty}^{V_r} \left(P_r - \frac{T_r}{V_r Z_{cr}} \right) dV_r + T_r \ln Z \\
&= g_A(T_r, V_r, \omega) \\
&= f_A(T_r, P_r, \omega)
\end{aligned}
\tag{2.18}
$$

In the approximation of (2.7) (and of the Law of the Corresponding States), therefore, all the fluids have the same departure function $A(T, V) - A^0(T, V^0)/RT_{cr}$ (except for the effects of the acentric factor ω).

Since the departure functions regarding entropy, internal energy and enthalpy derive from the departure function of the free energy of Helmholtz, generally speaking, it is also true

$$
\frac{H(T, V) - H^0(T, V^0)}{RT_{cr}} = f_H(T_r, P_r, \omega)
\tag{2.19a}
$$

$$
\frac{S(T, V) - S^0(T, V^0)}{R} = f_S(T_r, P_r, \omega)
\tag{2.19b}
$$

For the mono-component systems, as we know, at the liquid–vapour equilibrium the chemical potentials of the two phases are the same: $\mu_{l,sat}(T, P_{vp}) = \mu_{v,sat}(T, P_{vp})$. Since the chemical potential is $\mu = H - TS$, it is also true that $\mu_{l,sat}(T_r, P_{vp,r}) = \mu_{v,sat}(T_r, P_{vp,r})$. It follows that, within the limits of (2.7) (and of the Law of the Corresponding States), the relation between the reduced temperature T_r and the reduced vapour pressure $P_{vp,r}$ is a universal function for all pure fluids:

$$
P_{vp,r} = f_P(T_r, \omega)
\tag{2.20}
$$

Calculation results of the vapour pressure for certain fluids are shown in Fig. 2.3a. The fluids with the same value of acentric factor have, at the same T_r, a similar reduced vapour pressure. The observations above have a direct consequence on the evaporation heat and the shape of the limit curve.

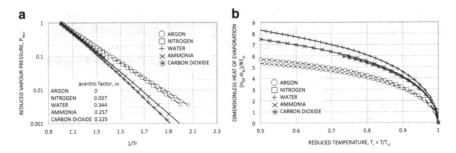

Fig. 2.3 (**a**) Reduced vapour pressure as a function of the reciprocal of the reduced temperature for certain fluids. (**b**) Dimensionless heat of evaporation as a function of the reduced temperature for some fluids. The results are those given by the equation of state (2.5a)

The Influence of the Nature of the Fluid on the Evaporation Heat

The evaporation heat can be derived directly from the difference between the departure functions calculated for the steam and the saturated liquid, using (2.19a), or by means of the Clapeyron equation:

$$\frac{dP_{vp}}{dT} = \frac{H_{SV} - H_{SL}}{T\,(V_{SV} - V_{SL})}$$

$$= \frac{H_{SV} - H_{SL}}{RT_{cr}} \frac{P_r}{T_r^2\,(Z_{SV} - Z_{SL})} \frac{P_{cr}}{T_{cr}} \tag{2.21}$$

since

$$\frac{dP_{vp}}{dT} = \frac{P_{cr}}{T_{cr}} \frac{dP_{vp,r}}{dT_r} = \frac{P_{cr}}{T_{cr}} \frac{df_P}{dT_r}$$

We obtain for the dimensionless molar evaporation heat, with regard to RT_{cr},

$$\frac{H_{SV} - H_{SL}}{RT_{cr}} = \frac{T_r^2\,(Z_{SV} - Z_{SL})}{P_r} \frac{df_P}{dT_r} \tag{2.22}$$

Therefore, the dimensionless evaporation heat is a universal function (except for the effects of the acentric factor ω) of just the reduced temperature (see Fig. 2.3b). In fact, it represents a particular enthalpic correction of the real gas: the correction, at constant pressure and temperature, corresponding to the phase change. The specific evaporation heat, once the reduced temperature is set, is more or less inversely proportional to the molar mass of the fluid. Several results are reported in Fig. 2.3b.

Fig. 2.4 Dimensionless
departure function for the
heat capacity C_P as a function
of the reduced pressure and
for several values of the
reduced temperature. The
results are relevant to
water [13]

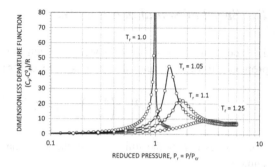

The Specific Heats of a Real Gas

The difference between the specific heats of a real gas under conditions T, V and the
corresponding values for the ideal gas can be determined directly from the equation
of state. In fact, since

$$\left(\frac{\partial S}{\partial T}\right)_V = \frac{C_V}{T} \quad \text{and}$$

$$C_P = C_V - \frac{T\,(\partial P/\partial T)_V^2}{(\partial P/\partial V)_T},$$

we obtain

$$C_V\,(T, V) - C_V^0\,(T) = T \int_\infty^V \left(\frac{\partial^2 P}{\partial T^2}\right)_V \mathrm{d}V \tag{2.23a}$$

$$C_P\,(T, V) - C_P^0\,(T) = T \int_\infty^V \left(\frac{\partial^2 P}{\partial T^2}\right)_V \mathrm{d}V - \frac{T\,(\partial P/\partial T)_V^2}{(\partial P/\partial V)_T} - R \tag{2.23b}$$

At the critical point $(\partial P/\partial V)_T$ is nullified and the departure function (2.23b)
tends to infinity. Consequently, there exists a zone of the thermodynamic diagram
inside of which there appear high and rapid variations in the specific heat. Figure 2.4
reports some results (for water) of the specific heat departure function. The
behaviour evident in the figure is common to all pure fluids. The dependence of the
heat capacity C_P on the pressure, as well as the temperature, introduces a difference
in heat capacity which, in regenerative gas cycles operating with working fluids that
have notable real gas effects, causes an irreversibility in the heat transmission to the
recuperators even with infinite surfaces (ideal recuperator).

The Speed of Sound in a Real Gas

The speed of sound v_s, or the speed at which the small perturbations of pressure propagate through a pure substance, is defined, for a single phase, as

$$v_s^2 = \left(\frac{\partial P}{\partial \rho}\right)_S = -V^2 \frac{C_P}{C_V}\left(\frac{\partial P}{\partial V}\right)_T \tag{2.24}$$

If the behaviour of the fluid is assimilable to that of a perfect gas, the speed of sound is

$$v_s^2 = \gamma(R/M)T \quad \text{or}$$

$$\frac{v_s^2}{(R/M)T_{cr}} = \gamma T_r \tag{2.25}$$

and, therefore, with the reduced temperature set, $v_s^2/(R/M)T_{cr}$ diminishes as the molecule complexity increases (or, with the increase in the number of atoms constituting the molecule, see Fig. 1.30) and v_s^2 diminishes with the molar mass of the gas.

If the gas is a real gas, the parameters which appear in (2.24) can be calculated by means of an equation of state, for example, using (2.5a). Then, within the limits of validity of the equation of state (and the Law of Corresponding States),

$$v_s^2 = -\frac{C_P}{C_V}P_{cr}V_{cr}k_1$$

$$= -\left[1 - \frac{k_2}{\frac{C_V^0}{Z_{cr}R} + k_3}\right]Z_{cr}(R/M)T_{cr}k_1 \quad \text{or}$$

$$\frac{v_s^2}{(R/M)T_{cr}} = -k_1 Z_{cr}\left[1 - \frac{k_2}{\frac{C_V^0}{Z_{cr}R} + k_3}\right] \tag{2.26}$$

with

$$k_1 = V_r^2 \left(\frac{\partial P_r}{\partial V_r}\right)_{T_r}$$

$$k_2 = \frac{T_r (\partial P_r/\partial T_r)_{V_r}^2}{(\partial P_r/\partial V_r)_{T_r}}$$

$$k_3 = T_r \int_\infty^{V_r}\left(\frac{\partial^2 P_r}{\partial T_r^2}\right)_{V_r} dV_r$$

Equation (2.26) shows that, with equal T_r and P_r, as the C_V^0/R increases, the ratio $v_s^2/(R/M)T_{cr}$ diminishes and if C_V^0/R tends to infinity, then $v_s^2/(R/M)T_{cr}$ tends to the value $(-k_1 Z_{cr})$.

The speed of sound is important, for example, in calculating the critical mass flow in nozzles [14]. Furthermore, the low speed of sound of gas and vapour, with complex molecules (consisting of numerous atoms) and high molar mass, could represent a significant limitation to fluid dynamics when designing turbine stages, with the possibility of shock waves and heavily penalised performance. The speed value v_s, as indicated in (2.26), is influenced decisively by the molecular complexity of the fluid, its molar mass, but also its volumetric behaviour. One example of an extreme condition caused by the effects of a real gas is the possibility that high values of the ratio C_V^0/R (around 50–70), associated with the intense effects of a real gas in proximity to the critical point of a fluid ($0.95 < T_r < 1.04$, $0.7 < P_r < 1.15$, see [15]), could introduce non-linear fluid dynamics, bringing about rarefaction shock waves and compression fans (the opposite of what would normally happen with an ideal gas). The study of the gas dynamics of heavy fluids near their critical point is known as "dense gas dynamics" and is a relatively modern field of study and research [16].

2.5 The Molecular Complexity of the Fluid and the Shape of the Limit Curve

The physical quantity which is most directly correlated to the molecular complexity of a fluid, as we have seen, is its specific molar heat (C_P or C_V, which are strictly correlated). The slope of the saturated vapour line can be calculated and a σ parameter introduced which, in its turn, can be assumed to be a parameter linked to the molecular complexity of the fluid [17]

$$
\begin{aligned}
\sigma &= \frac{T_{cr}}{R} \left[\frac{dS_{SV}}{dT} \right]_{T_r=0.7} \\
&= \frac{T_{cr}}{R} \left[\left(\frac{\partial S}{\partial T} \right)_P + \left(\frac{\partial S}{\partial P} \right)_T \frac{dP_{vp}}{dT} \right]_{SV,T_r=0.7} \\
&= \frac{T_{cr}}{R} \left[\frac{C_P}{T} - \left(\frac{\partial V}{\partial T} \right)_P \frac{dP_{vp}}{dT} \right]_{SV,T_r=0.7}
\end{aligned}
\tag{2.27}
$$

For a real gas, the σ parameter is calculated by means of an equation of state (for example, using (2.5a)). The σ coefficient is primarily a function of the heat capacity of the saturated vapour and, as a consequence, is directly correlated to the molecular structure of the fluid. The qualitative effects of the fluid's molecular structure on the σ value can easily be shown if the vapour is assimilated to an ideal gas. In which case,

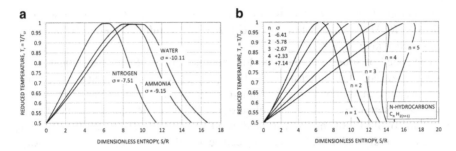

Fig. 2.5 Tendency of the saturation dome in the plane of reduced temperature-reduced entropy for various fluids. The results are those given by the equation of state (2.5a). (**a**) Typical fluids with simple molecular structure. (**b**) First compounds in the series of aliphatic alkane hydrocarbons

$$\sigma \approx \frac{T_{cr}}{R} \left[\frac{C_P^0}{T} - \frac{R}{P} \frac{dP_{vp}}{dT} \right]_{T_r=0.7}$$

$$= \left[\frac{\gamma}{\gamma - 1} \frac{1}{T_r} - \frac{1}{P_r} \frac{df_P}{dT_r} \right]_{T_r=0.7} \qquad (2.28)$$

The slope of the upper limit curve, in the thermodynamic plane T-S, may then be negative, null or positive according to which of the two terms included within square brackets prevails. The fluids with a complex molecular structure, i.e. formed by a large number of atoms, tend to have a high specific molar heat and the ratio $\gamma = C_P/C_V$ tends to unity (see Fig. 1.30). Consequently, they are characterised by a slope of the upper limit curve that is markedly positive. Fluids with a simple molecular structure, like water and ammonia, have modest specific heat values and, therefore, the σ is negative for them. Figure 2.5 shows several examples of limit curves for different fluids.

The vapour pressure for any pure fluid may be expressed by means of an equation such as (see (2.20) and Fig. 2.3a):

$$\log P_{vp,r} = A + \frac{B}{T_r}$$

Using the definition (2.6) of the acentric factor and the condition $P_r = 1$ when $T_r = 1$, the two parameters A and B can be calculated and the vapour pressure curve can be expressed as

$$\log P_{vp,r} = \frac{7}{3} (\omega + 1) \left(1 - \frac{1}{T_r} \right)$$

Fig. 2.6 Parameter of molecular complexity sigma, evaluated by means of (2.29), as a function of the ratio C_P^0/R for some pure fluids

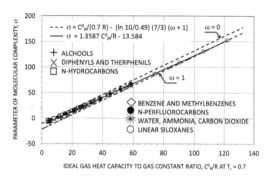

which, substituted in (2.28), gives for the σ parameter

$$\sigma \approx \frac{C_P^0}{0.7R} - \frac{\ln 10}{0.49} \times \frac{7}{3}(\omega + 1) \qquad (2.29)$$

The slope of the upper limit curve is positive if $\sigma \geq 0$ or if C_P^0/R is higher than the value $7.67(\omega + 1)$. Figure 2.6 reports the σ parameter, calculated for several fluids by means of (2.29). The effect of the acentric factor on the σ value is modest and, generally speaking, σ is primarily a function of the ratio C_P^0/R, which changes 30 times in passing from the more simple molecule structures (water, ammonia, carbon dioxide) to those which are more complex (like, for example, the linear siloxanes).

As we shall see in Chap. 3, the shape of the limit curve, its slope in particular, is important because, of the numerous thermo-physical properties that characterise each fluid, only a few play a significant role in determining the "quality" of a thermodynamic cycle on the limit curve. Generally speaking, two properties, above all, are responsible for both the cycle performance and the degree of suitability of the fluid for heat recovery from a determined heat source: the critical temperature T_{cr} and the parameter of molecular complexity σ. In general, at least for homologous fluids, an increase in the molecular complexity is accompanied by a decrease in the critical pressure and a rise in the critical temperature of the fluid.

2.6 The Equations of State for Mixtures

For the multicomponent mixtures in a single phase, it is usual to define the volumetric equation of state (2.2), starting with a valid equation for pure fluids and introducing appropriate terms which, correlating between them the valid parameters for pure fluids, provide (via the so-called mixing rules) the corresponding parameters of the mixture.

In the case of (2.5), the mixture parameters are defined, for example, by the following mixing rules [5]:

$$a = \sum_i \sum_j y_i y_j a_{ij} \tag{2.30a}$$

$$b = \sum_i y_i b_i \quad \text{where} \tag{2.30b}$$

$$a_{ij} = \left(1 - \delta_{ij}\right) a_i^{1/2} a_j^{1/2} \quad i \neq j \tag{2.30c}$$

and $a_{ii} = a_i$, $a_{jj} = a_j$, $\delta_{ii} = \delta_{jj} = 0$.

The coefficient δ_{ij} is the specific "binary interaction coefficient" of the couple $i - j$. The binary interaction parameters are often estimated on the basis of experimental data for the liquid–vapour equilibrium relative to the mixture.

The departure functions defined for the pure fluid are also valid for the mixtures, provided that the parameters a and b of the mixture are used with its specific composition. The specific heats under conditions of an ideal gas for the mixture should be calculated using the following equations:

$$C_V^0 = \sum_i y_i C_{V,i}^0 \tag{2.31a}$$

$$C_P^0 = \sum_i y_i C_{P,i}^0 \tag{2.31b}$$

or as weighted mean with the molar fractions of the pure components of the mixture. A critical point can be defined and calculated also for the mixtures, applying the appropriate conditions of stability [4, p. 6.30], [12, Chap. 7].

Exercises

2.1. Figure 2.7 reports the liquid–vapour equilibrium pressures for pure ethane (fluid number 1), for pure n-heptane (fluid number 2) and for an equimolar mixture of the two ($y_1 = y_2 = 0.5$). The figure also shows the critical point locus, that is, the variations in pressure and critical temperature with the composition of the mixture (the experimental results are taken from [18]). Calculations of the liquid–vapour equilibrium were made using the equation of state (2.2), with the mixing rules (2.30), assuming $\delta_{12} = \delta_{21} = 6.7 \times 10^{-3}$.

As can be seen from the figure, mixing the two fluids, the pressure and the critical temperature vary with continuity from the values relative to the case of pure ethane ($y_1 = 1$) to the values of pure n-heptane ($y_1 = 0$). These variations, though, are not linear: firstly, the critical pressure increases, reaching its maximum value of around 88 bar in correspondence to $y_1 \approx 0.78$, then dropping to a minimum value that

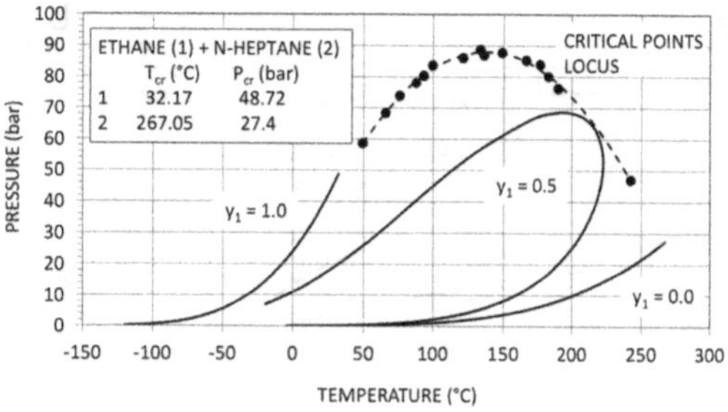

Fig. 2.7 Bubble and dew
pressure curves per ethane,
per n-heptane and for a mix
of the two

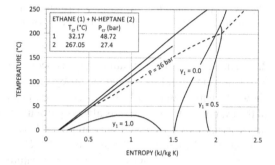

Fig. 2.8 Trend of the limit
curve in the thermodynamic
plane T-S for pure ethane, for
pure n-heptane and for a
mixture of the two. For the
mixture, there is also shown
the isobar at 26 bar

corresponds to the case of pure n-heptane. The mixing of two (or more) compounds
represents, therefore, an efficient instrument for varying the critical point of a
fluid.

As can be seen, though, in the case of the equimolar mixture, the mixture, once
the pressure has been fixed, is characterised by a difference between the bubble and
the dew temperatures (the so-called temperature glide). For example, in this specific
case, at the pressure of 26 bar, the bubble point is reached at the temperature of
about 50 °C and the corresponding dew point is around 200 °C. The temperature
glide diminishes as the pressure increases.

Figure 2.8 shows the limit curves in the thermodynamic plane T-S. As we can see,
as the composition varies, as well as the pressure and the temperature at the critical
point, there is also a significant change in the molecular complexity (the slope of
the upper limit curve changes). The figure also clearly shows the temperature glide
associated with the isobar of 26 bar.

The results shown here for the mixture of ethane and n-heptane are basically and
qualitatively indicative of the thermodynamic behaviour of all the mixtures known
as non-azeotropic.[5]

---

[5] An azeotrope is a mixture of two or more substances with a composition that cannot be distilled.
That is, the glide temperature at constant pressure is null and the mixture behaves, from this point of
view, as if it were a pure fluid. The composition of the liquid phase is the same of the composition
of the vapour phase.

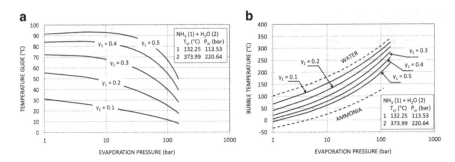

Fig. 2.9 Temperature glide and bubble temperature for ammonia–water mixtures, with various compositions. The results are those given by the equation of state (2.5a) $\delta_{12} = -0.2589$

2.2. The mixture of water–ammonia has been proposed—and occasionally used—as the working fluid in special steam cycles (see, for example [19]). Figure 2.9a shows the temperature glide, for different compositions of the mixture, as a function of the evaporation pressure: it is always particularly high and grows rapidly with the molar fraction of ammonia. For example, at 1 bar of pressure, the temperature glide is about 31 °C when $y_1 = 0.1$ and about 92 °C when $y_1 = 0.5$.

Having set the composition, the temperature glide falls as the evaporation pressure increases. For example, when $y_1 = 0.3$, we get 72 °C at the evaporation pressure of 1 bar and 29 °C at the pressure of 150 bar. To get significant temperature glide at high pressure (advantageous, in principle, if the mixture is used in heat engines that recover heat from heat sources with variable temperature), we need mixtures that are sufficiently rich in ammonia; modest temperature glide in condensation (not excessively penalising from a thermodynamic point of view) is obtained, though, when the molar fraction of ammonia is small.

Figure 2.9b shows the bubble temperature as a function of the evaporation pressure for mixtures of water and ammonia, with different compositions. Bubble temperature values close to the environmental values are obtained when $y_1 \approx 0.3$, which could represent the composition of the base mixture in a thermodynamic cycle that uses a water–ammonia mixture as working fluid.

References

1. Smith JM, van Ness HC (1975) Introduction to chemical engineering thermodynamics, 3rd edn. McGraw-Hill Kogakusha, Tokyo
2. Gyftopoulos EP, Beretta GP (2005) Thermodynamics. Foundations and applications. Dover, Mineola, NY
3. Deiters UK, De Reuck KM (1999) Guidelines for publication of equations of state I. Fluid fluids. Fluid Phase Equil 161:205–219
4. Poling B, Prausnitz JM, O'Connell JP (2001) The properties of gases and liquids, 5th edn. McGraw-Hill, New York

5. Peng D-Y, Robinson DB (1976) A new two-constant equation of state. Ind Eng Chem Fund 15(1):59–64
6. Pitzer KS, Lippmann DZ, Curl RF, Huggins CM, Petersen DE (1955) The volumetric and thermodynamic properties of fluids. II. Compressibility factor, vapour pressure and entropy of vaporization. J Am Chem Soc 77(13):3433–3440
7. Hougen OA, Watson KM, Ragatz RA (1959) Chemical process principle. Part II – thermodynamics, 2nd edn. Wiley, New York, NY
8. Hirschfelder JO, Buehler RJ, McGee HA, Sutton JR (1958) Generalized equation of state for gases and liquids. Ind Eng Chem 50(3):375–385
9. Schreiber DR, Pitzer KS (1988) Selected equation of state in the acentric factor system. Int J Thermophys 9(6):965–974
10. Reynolds WC (1979) Thermodynamic properties in SI. Graphs, tables, and computational equations for forty substances. Department of Mechanical Engineering, Standford University, Stanford, CA
11. Gross J, Sadowski G (2001) Perturbed-chain SAFT: an equation of state based on a perturbation theory for chain molecules. Ind Eng Chem Res 40:1244–1260
12. Tester JW, Modell M (1996) Thermodynamics and its applications, 3rd edn. Prentice Hall PTR, Upper Saddle River, NJ
13. Parry WT, Bellows JC, Gallagher JS, Harvey AH (2000) ASME international steam tables for industrial use. CRTD-Vol. 58. ASME, New York, NY
14. Johnson RC (1964) Calculations of real-gas effects in flow through critical-flow nozzles. ASME J Basic Eng 86(3):519–526
15. Angelino G (1973) About the possibility of the occurrence of rarefaction shock waves in fluids. La Termotecnica 27(9):489–494
16. Brown BP, Argrow BM (2000) Application of Bethe-Zel'dovich-Thompson fluids in organic rankine cycle engines. J Propul Power 16(6):1118–1124
17. Invernizzi C, Iora P, Silva P (2007) Bottoming micro-rankine cycles for micro-gas turbines. Appl Therm Eng 27(1):100–110
18. Hicks CP, Young CL (1975) The gas-liquid critical properties of binary mixtures. Chem Rev 75(2):119–175
19. Ibrahim OM, Klein SA (1996) Absorption power cycles. Energy 21(1):21–27
20. Casci C (1978) Elements of fluid machines - two phase fluid machines. Masson Italia Editori, Milan (in Italian)

Chapter 3
The Organic Rankine Cycle

As already discussed in Sect. 1.6, the energy conversion plants based on the steam Rankine cycle guarantee high levels of efficiency (40–50 %) and are characterised by good thermodynamic qualities in cases where the useful power is high (generally, we are talking of hundreds of MWe), so as to justify the costs of the plant incurred by the elevated cycle pressure (150–300 bar), the high maximum temperatures (500–700 °C), all the complexity of the plant layout (numerous regenerators) and of the primary movers (turbine expander).

The current industrial interest in energy saving, in the use—where possible—of renewable energy and in heat recovery, necessitates the use of highly efficient thermodynamic tools: these should be suitable for the variety of heat sources available and for varying sizes of power supply.

For example, a significant quantity of waste heat is produced by industrial processes, at relatively high temperatures (200–600 °C), frequently available as the sensible heat of gas flows or from vapour condensation (of pure or multicomponent vapours). In such cases, the steam may be thermodynamically and economically inadequate.

In Italy, for instance, a recent study has estimated that, just for the waste heat recovery using Organic Rankine Cycles (ORC) technology of the size of 0.5–5 MW for the sectors of metalworking, cementmaking and glassmaking, there is a potential installed power of around 130 MW, corresponding to a total of between 640 and 1,025 GWh of producible electricity per year.

By way of a further example, in the sector of renewable energy, several forecasts agree on the enormous potential for exploiting the geothermal energy resources of the planet. The geothermal resources often consist of multicomponent fluids and vapours, also with significant quantities of incondensable gases, which could render them both difficult and inappropriate for use directly in steam cycles. In such cases, the use of organic fluid engines could be the only valid alternative.

In general, therefore, wherever the use of steam is unsatisfactory due to low efficiency levels or for the excessive plant costs, systems based on the use of Rankine cycles that use working fluids other than steam could well be the answer.

C.M. Invernizzi, *Closed Power Cycles*, Lecture Notes in Energy 11,
DOI 10.1007/978-1-4471-5140-1_3, © Springer-Verlag London 2013

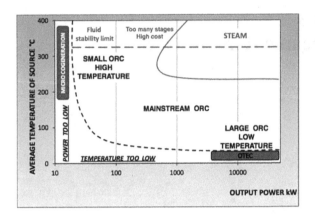

Fig. 3.1 Fields of employment for the Rankine cycle engine: steam cycles and Organic Rankine Cycles (from author [1])

Such engines have been developed and are still the object of intense research and development projects. The fluids used tend to be organic fluids and the thermodynamic cycles that utilise them are called ORC.

For special applications, the organic fluids give indubitable advantages over the use of steam, but, on the downside, they are sometimes costly, are characterised by poor thermal exchange properties and involve potential risks to safety (inflammability) and health (toxicity).

Figure 3.1 represents in a synthetic but approximate manner the field of employment, in terms of temperature and power, for heat engines using both steam and organic fluid. Steam has practically no rivals in applications requiring great power levels (higher than 500–1,000 kW) and at medium–high temperatures (above 200–250 °C). For lower power levels, the steam turbine is usually too expensive. Organic fluids have a heat stability limit that restricts their use to maximum temperatures of 300–400 °C. Temperatures that are excessively low (below 70–100 °C) render the costs of the heat engine so high that it is often inadvisable to use them. Unless, that is, the power levels are particularly high (as in the case of OTEC systems). Very low power levels (below a few dozen kWe, typical in micro-cogeneration systems) mean high costs for the organic fluid engine and, consequently, it becomes at present often uneconomic.

In any case, the range of potential applications for the ORC, at low and medium–high temperatures, is vast and interest in the technology is growing rapidly. Growth potential is particularly strong in all those sectors of primary generation and cogeneration (including the domestic field, within the range of a few kW); in systems using biomass (both residual and in the form of energy crops), when there are difficult fuels involved (syngas, flare gas, etc.); in the solar thermodynamic sector; in the numerous cases of heat recovery (from industrial processes, from gas turbines, from reciprocating ICE, in plants for re-gasification of natural gas, etc.) and in the exploitation of geothermal sources and in OTEC systems.

There are now commercially available turbomachines or expanders for ORC with power levels below 50 kW and ever smaller organic fluid engines are being proposed with scroll expanders. The greatest challenge for the industry of ORC systems today, especially for cases of low temperature and low power levels, is not so much the thermodynamic efficiency, which is potentially already very good, but the reduction of costs. In this sense, the synergy that now exists between the HVAC sector and the companies that develop ORC systems appears very promising, although the air-conditioning industry has developed and continues to develop refrigerating fluids that sometimes appear—from a strictly thermodynamic point of view—far from optimal for heat engines.

In this chapter, we shall discuss the principal characteristics of organic fluids and the Rankine cycles that can be made using them, with some real examples of engines.

3.1 A Brief History of the Organic Rankine Engines

The ORC, as mentioned above, is a Rankine cycle that does not use steam as its working fluid, but another fluid, generally an organic fluid.[1]

William J.M. Rankine developed a complete theory of the steam engine in his famous manual of 1859 [2], but, as early as 1825–1826, Thomas Howard had made an engine using "alcohol" or "ether" as working fluid. Various authors refer to it and describe it ([3, p. 217], [4, p. 599], [5, p. 644]). In [5] it says

> The intention of the inventor of this engine, was the employment of alcohol or ether, as a motive power, on the ground of their exerting a much greater expansive force than steam, at similar temperatures.

It appears that the engine, with a design power of 24 hp (about 18 kW), worked for a brief period in Rotherhithe, Surrey, UK.

The British journal The Engineer, on 9 January 1885,[2] contains an article that describes the results of a test carried out by a commission of the United States naval engineers on a launch engine, which used as its working fluid a solution of water and methyl alcohol[3] (with 5–15 % methyl alcohol): a binary vapour engine.[4]

---

[1]By organic fluid or compost we mean any compost, not necessarily present in living organisms, containing a significant quantity of carbon.

[2]A great deal of interesting information and anecdotes about unusual working fluids and much else can be found in "The Museum of Retro Technology". Available at http://www.douglas-self.com/MUSEUM/museum.htm. (cited May 27, 2012)

[3]The methyl alcohol, known also as methanol, is a well-known alcohol fuel. Its chemical formula is CH_3OH, its boiling point is 65 °C and at 25 °C it has a vapour pressure of 0.17 bar. The critical temperature is 239 °C and the critical pressure 81 bar.

[4]Here, the term "binary" refers to the working fluid, obtained by mixing two compounds. In its modern usage, the term "binary" identifies an engine with two working fluids that are distinct and physically separate from each other, each operating at a different temperature.

The commission made a rigorous comparison with the performance of the same engine when using steam. This showed, for the binary vapour engine, a specific fuel consumption of 5.07 lb of coal per hourly horsepower, compared to a specific consumption of 5.76 lb of coal per hourly horsepower in the case of the steam engine. It is worth reporting the conclusions of the commission:

> although it is not clear in the minds of the Board how a vapour requiring less heat in its production can give up more of its heat in the production of power than another vapour under like conditions containing a greater quantity of work, yet, accepting the result of trials as being absolutely correct, it would be well to see what this saving costs.

Given the market price of methyl alcohol and the excessive leakage of fluid from the engine, the economic results were decidedly unfavourable for the binary vapour engine. A subsequent engine, of greater power (150 hp, compared to the 10 hp of the prototype), did not work at all, due to the distillation of methyl alcohol in the boiler and the uncontrollable leakage of nauseating fumes.

Around 1850, Du Tremblay, an engineer from Lyon, created a binary heat engine with steam for the engine at high temperature and with "ether" (probably diethyl ether,[5] a compound that is much more volatile than water) for the engine at a lower temperature [6, p. 131]. After evaporation in the boiler and expansion in the cylinders, the steam released its thermal energy of condensation to the second engine, causing the evaporation of the ether which, expanding in another cylinder, produced further work. In this way, he created a real binary cycle (see Chap. 5).

The engine was installed in several passenger ships, but, following an explosion caused by the ether, at the port of Bahia, South America, in 1856, the construction of steam-ether binary motors was interrupted.

Towards the end of the nineteenth century, some small boats were made with engines that used boiling petrol in place of steam for the power system. The most wellknown of these were those built by Frank W. Ofeldt (the "naphtha launches"), which entered production from 1897. The working fluid for the engine was naphtha,[6] which was also used as lubricant for the moving parts and as fuel for the evaporation of the working fluid. These launches enjoyed a moderate commercial success thanks to the existence, in the United States, of a law obliging the presence of a qualified engineer aboard whenever there was a steam engine present. The law did not apply to boilers using other types of vapours.

However, it was not until the twentieth century that the ORC was developed intensively. In the sector of thermodynamic conversion of solar energy, Tito Romagnoli, an Italian, developed several Rankine engines between 1923 and

---

[5]The chemical formula for diethylether, which is extremely inflammable, is $CH_3–CH_2–O–CH_2–CH_3$, and its boiling point is 34 °C; at a temperature of 25 °C, the vapour pressure is 0.71 bar. The critical temperature is 194 °C and the critical pressure 36.4 bar.

[6]The term "naphtha" is generally used to identify mixtures of hydrocarbons. Naphtha is often used as a feedstock for the production of gasoline.

1930: one, around 2 hp (about 1.5 kW), with methyl chloride[7] as its working fluid. He used fixed solar collectors, probably flat, without concentration [7, 8].

Between 1961 and 1962, Harry Zvi Tabor and Lucien Bronicki built various Rankine engines with monochlorobenzene[8] at 140–150 °C, with power from 2–10 kW [9].

Some solar pumping systems were developed, in the years 1977–1978, with engines using refrigerant 113[9] from 20–40 kW [9, 10, p. 639]. Solar Rankine engines using toluene,[10], with parabolic dish collectors, and a maximum temperature of 400 °C, were made between 1983 and 1984 with power from 25–100 kW.

In 1967, in the Soviet Union on the Kamchatka peninsula, what is generally considered to be the first geothermal binary cycle was installed. The working fluid was the refrigerant 12[11] and the engine had a gross power of 680 kW. The heat source was geothermal water at low temperature (80 °C) [11, p. 519].

The ORC started to be taken into serious consideration in the late 1980s, also for use as devices for the dynamic conversion of solar energy in space, since, unlike the closed Brayton cycles; by functioning at lower maximum temperatures (400 °C compared to the 750–800 °C necessary for closed gas cycles, see Sect. 1.7), they enabled the use of lighter and more conservatively designed concentrators, guaranteeing high performance at concentration ratios below 500, and, at the same time, allowing numerous materials to be employed in the heat storage [12].

By 1981, there were 2,150 Rankine engines with 16 different working fluids, made by about 20 different engine manufacturers. In terms of the number of engines and operating time, the leading manufacturer of Organic Rankine engines at the time had made 2,000 of them, with a total of 18,000,000 operating hours, using trichlorobenzene as the working fluid (see [13]).

Starting with the early attempts (at the end of the nineteenth century) and the organic fluid engines that were built up to the 1970s and 1980s, there have been numerous models. Far more than those briefly listed above: the references [7, 9, 13] give a fairly complete description of them. This brief overview, though, makes it clear how the initial fluids employed were those of the refrigeration industry: not only the organic fluids in the strict sense, but also sulphur dioxide (Henry E. Willsie, in 1904, built two solar power plants with SO_2 1 of 6 hp and 1 of 15 hp) and

---

[7]The methyl chloride, or chloromethane, with a chemical formula CH_3Cl, has a boiling point of −24.2 °C.

[8]Chemical formula C_6H_5Cl, boiling point 131.76 °C, critical temperature 359.25 °C and critical pressure 45.2 bar

[9]1,1,2-trichloro-1,2.2-trifluoroethane, with boiling point of 47.7 °C

[10]Toluene, or methylbenzene, with chemical formula $C_6H_5CH_3$. Boiling point at 110.4 °C, critical temperature 318.6 °C and critical pressure 41.08 bar

[11]Dichlorodifluoromethane, CCl_2F_2. Boiling point −29.8 °C.

ammonia (for example, the Campbell engine, in 1891). In the first machines, the expander used was mainly volumetric; the turbine was introduced later. There have been numerous solar applications, sometimes resorting to concentration collectors, mostly flat collectors, at low temperatures and with modest power levels. In some cases, the natural salt gradient of the so-called solar ponds was exploited.[12]

The Italian School

Italy has a long tradition in the development of heat engines based on ORC. The studies and the models of the engineer Romagnoli (between 1923 and 1931) have already been mentioned. It is also worth recording those proposed by Prof. Mario Dornig[13] of the Institute of Technical Physics at the Polytechnic of Milan, in the years 1918–1922 (see [14]) and, subsequently, those of Prof. Luigi d'Amelio,[14] professor at what was then the Regio Istituto Superiore d'Ingegneria of Naples, who (see [8]) dedicating himself to the use of solar energy for raising water for agricultural uses in Libya, made detailed studies of an engine using vapours of ethyl chloride.[15] The vapour expansion was ensured by an impulse single-stage turbine, and the working fluid chosen was methyl chloride because, among other things, it meant that the rotor could be made with a peripheral speed lower than 100 m/s, which was then considered to be the prudent top speed. The temperatures of evaporation and condensation were 40 °C and 23 °C, respectively, with a turbine shaft power of around 6 hp (4.5 kW). On the basis of these studies, he then built a pilot plant of 11 kW, in 1940, on the Island of Ischia, where it could use the heat of the local thermal springs. The results were good; however, a second plant of 250 kW, finished in 1943, never went into operation.

A solar engine, at low temperature (about 100 °C) and operating with sulphur dioxide, was designed in the early 1930s by Daniele Gasperini (1895–1960), an

---

[12]A solar pond is a vast area of salt water which, due to the favourable salt gradient, behaves as a large flat solar collector of thermal energy. A solar pond can be used for a variety of applications, amongst which is the generation of electricity.

[13]Born in Florence in 1880 to parents originating from Trieste. Having graduated in Civil Engineering at Rome in 1904, he moved to Munich, where, in 1911, he obtained his doctorate in Thermal and Mechanical Science. Returning to Italy, in 1917 he obtained the professorship in Fluid Machinery at the Polytechnic of Milan, which he held until his retirement. (1951). He died on 12 November 1962.

[14]Born in Naples on 1 June 1893, he was a professor at the University of Naples and headed the Institute of Thermal, Hydraulic and Agricultural Machinery until 1963. He died in Naples on 1 December 1967.

[15]Ethyl chloride, or chloroethane, has the chemical formula CH_3-CH_2Cl, a boiling point of 12.35 °C, a critical temperature of 187.25 °C and a critical pressure of 52.7 bar.

artisan and refrigeration expert from Rovereto (Trento).[16] During his work in Libya, he collaborated with Giovanni Andri to build the first engine model, exhibited at the fair of Tripoli in July 1936. After the Second World War, Gasperini continued to develop his idea and, with the collaboration of Ferruccio Grassi (1897–1980), an engineer from Lecco, he designed and built a solar pump for raising water from below ground, called SOMOR, after the name of the company that built it (SOcietá MOtori Recupero—Company of Recovery Engines—for solar heat and waste heat), exhibited at the first world fair on solar energy, held at Phoenix, Arizona, in 1955 [15].

However, the real Italian School, which significantly and systematically contributed to the development and study of Rankine engines using organic fluid appeared during the second half of the 1960s at the Polytechnic of Milan. Its founder was Prof. Gianfranco Angelino,[17] with Prof. Mario Gaia and Prof. Ennio Macchi, first as his pupils, then as colleagues and friends.

Between 1976 and 1984, the Polytechnic group designed and contributed, with the financial backing of various institutions and private companies, to the realisation of 14 ORC engines, from 3 to 500 kW, to use various heat sources (solar energy, geothermal fluids, industrial waste heat, fossil fuels) [16]. Among these, we should note the following for their peculiarity:

- Several small engines using perchloroethylene[18] (from 3 to 12 kW) at low temperature (from 75 °C to 83 °C, with corresponding evaporation pressures of 0.235–0.306 bar), in which the working fluid was chosen in order to give good turbine design.

---

[16]In the newspaper The Deseret News—November 15, 1951 (the oldest daily newspaper published in the state of Utah, at Salt Lake City)—there can be read the following brief but curious note:

The Sun Could Supply Electricity, by A. De Montmorency New York, November 14—A new Italian invention will permit each house to generate its own electricity without any expense of fuel, simply by using the sun's energy. A dispatch from Milan to informations of Madrid reported that Prof. Mario Dorning of that Lombardian city had built with the help of Daniel Gasperini, an engineer, a solar engine capable of producing 10 kilowatt-hours daily. Three such machines have been sent to Egypt for a tryout.

[17]Born in Naples on 18 October 1938. He graduated with full honours in 1962, discussing his thesis on "Prestazioni degli effusori a spina nei propulsori a razzo". In the same year that he graduated, he won an AGARD scholarship for a specialisation course in Experimental Aerodynamics at the Centre de Formation en Aérodynamique Expérimentale, Von Kármán Institute, in Belgium. In 1963 he was awarded his diploma with distinction and received the Theodore Von Kármán prize, which was reserved for the best student of each year. From 1973 to 2009 he was full-time professor in Machinery at the Polytechnic of Milan. He was director of the Machinery Section at the Department of Energy at the Polytechnic of Milan and, for many years, director of the research doctorate in Energy. He died on 9 May 2010.

[18]Perchloroethylene, or tetrachloroethylene, $Cl_2C=CCl_2$ is an excellent solvent of organic substances, which is not particularly volatile and is non-inflammable. For these reasons, it is widely used these days in dry cleaning. It has a boiling point of 121.1 °C

- An engine of 500 kW with refrigerant 113,[19] in which a working fluid was chosen that had a critical temperature and molecular complexity that would guarantee an efficient cooling of the heat source (a geothermal fluid).
- Two high temperature (280 °C and 340 °C) perfluorocarbon engines,[20] of 45 and 25 kW, designed to give high efficiency, with a multistage turbine and a demanding regenerator.
- A cogeneration engine with dichlorobenzene,[21] with condensation temperature at 80 °C (condensation pressure 3.5 kPa) and of 100 kW with a 3,000 rpm turbine and a direct-drive turbo-alternator.

Up until the end of the 1980s, the organic fluid engines were prototopical machines, or almost, with an extremely limited market,[22] whilst, today, the use of ORC is expanding rapidly and the basic technology is wellknown. The most common applications have been in the biomass and geothermal sectors, whilst the biggest margin for growth is forecast in the fields of heat recovery and solar thermodynamics (also, in principle, in OTEC plants). In Europe alone, there are now around 200–230 plants operating with organic fluid engines: 80 in Germany, around 70 in Italy and 30 in Austria.

This brief historical overview of the use of organic fluids as working fluids in the energy conversion sector would not be complete without citing the profound studies carried out between the end of the 1950s and the early 1970s into the use of organic fluids in nuclear reactors: the organic-cooled and moderated reactors (OCMR). The abundance of hydrogen in the molecule of organic fluid guarantees excellent moderating capabilities (comparable to those of water) but with a series of advantages for the organic refrigerants with respect to water: vapour pressure is very low at high temperatures, excellent compatibility with carbon steel and low induced radioactivity. Furthermore, their physical properties are wellknown and the cost of the fluids relatively modest. If heavy water were used as the moderator, then, in principle, even natural uranium could be used as fuel. On the downside, the decomposition induced by the temperature and the radiation, together with the relatively low heat exchange capabilities and the inflammability, represent important negative characteristics that discourage the practical use of organic fluids as coolants

---

[19]1,2-Dichlorotetrafluoroethane, $ClF_2C–CF_2Cl$, boiling point 3.43 °C, critical temperature 145.75 °C, critical pressure 32.37 bar.

[20]One with Flutec PP3, perfluoro-1,3-dimethylcyclohexane, C_8F_{16}, boiling point 102 °C, critical temperature 241.55 °C and vapour pressure 4.8 kPa at 25 °C; the second with Flutec PP5, perfluorodecalin, or perfluoronaphtalene, $C_{10}F_{18}$, boiling point 142 °C, critical temperature 292 °C and vapour pressure 0.88 kPa at 25 °C

[21]1,4-Dichlorobenzene, $C_6H_4Cl_2$, boiling point 174 °C

[22]A notable exception was (and continues to be after 40 years) a small generator for generating energy in remote places, called the Remote Power Unit: made with power levels between 600 W and 4 kW, it is very widespread.

in nuclear reactors. The fluids under study were various polyphenyls,[23] with high boiling points (300–400 °C) and critical temperatures of around 600 °C. Besides the research carried out in the United States, similar studies were undertaken in Canada and in the ex-Soviet Union and there was even a detailed Italian project (Progetto Reattore Organico) between 1960 and 1970.

3.2   The Characteristics of Candidate Working Fluids According to the Applications

Engines operating according to a Rankine cycle, with an organic working fluid, generally with a high molecular mass, present characteristics that make them extremely interesting for applications at medium-low temperatures and relatively low power levels (maximum power of a single engine in the order of several MW, at most). The reasons behind the use of fluids other than steam in Rankine cycles are, primarily, thermodynamic and linked to a well-designed expander, usually a turbine. In fact, the choice of an appropriate working fluid will satisfy a variety of needs:

- Fluids with different critical parameters (temperature and pressure) assure that configurations of the thermodynamic cycle become possible to be inaccessible in the state diagram of water, for instance, supercritical cycles even at low maximum temperatures.
- Even where there are great ratios between the temperature of the heat source and the temperature of the cold well, efficient thermodynamic cycles can be achieved with a relatively simple plant set-up and perhaps with one expansion stage, thanks to the regeneration that occurs with a de-superheating of the vapour at the turbine outlet and without having recourse to vapour extraction.
- The turbine requires, for the most part, modest peripheral speed and condensation is avoided during the expansion. The turbine, though, often has supersonic flows with high expansion ratios.
- The choice of fluids, influencing the volumetric flows, permits turbine optimisation for any power level.
- The pressure levels (and the expansion ratios) between the various components may be chosen with a certain freedom, independently of the temperature levels of the heat source and the cold well (for example, low temperatures may be associated with high pressures and high temperatures with low pressures).

The final choice of the working fluid must, inevitably, be influenced by safety and financial considerations. Thus, it is generally the result of a reasonable compromise between the necessary and the desirable characteristics of working fluids in

---

[23]*Ortho-*, *meta-* and *para*-terphenyls ($(C_6H_5)_2C_6H_4$) in mixtures. A prototype of a commercial reactor, of 45.5 MWt, was made anyway (under a project of 1956 and the reactor was operational from 1963 to 1966) in Ohio in the USA: the Piqua OMR plant [17].

Rankine power systems: primarily, adequate thermo-physical and thermodynamic properties, compatibility with the materials and the limits of thermal stability of the fluid, the health and safety characteristics, the fluid's availability and its cost. The thermo-physical properties and characteristics of the fluid and their effects upon the performance of the thermodynamic cycle will be discussed in the next section. Here, briefly, we consider the other aspects.

Material Compatibility and Thermochemical Stability Limits

Although organic fluid engines, when designed well, are characterised by an optimal thermodynamic quality, their performance in absolute terms depends on the maximum operating temperatures (see Sect. 1.1).

The choice of the most suitable working fluid passes therefore, necessarily, after an analysis of its thermodynamic behaviour, through the study of its thermal stability and its chemical compatibility, both at room temperature and at operating temperatures, with the materials used in building the plant: those materials used in making the turbines, the heat exchangers and the pumps and used for the tubes, for the seals and for the lubricants. The fluid should be thermally and chemically stable at all operating temperatures, not just in the presence of materials commonly used, but also in the presence of air and water. In any case, the decomposition, which is often inevitable, must occur at a sufficiently slow rate as not to compromise the fluid characteristics and deteriorate excessively the thermodynamic performance of the system.

Tests of various natures have been carried out to establish values of reasonable safety for the maximum operating temperatures. Sometimes this has been done using containing materials for the fluid that are as inert as possible, with the aim of determining the intrinsic molecular stability, sometimes by choosing the materials a priori in order to evaluate their compatibility with the fluid, rather than to investigate the absolute stability. From time to time, different physical properties have been assumed, more or less arbitrarily, as indicators of decomposition in progress (isothermal increases in pressure, formation of carbonaceous deposits, etc.). Catalysis and corrosion can have significant effects on experimental results. Ideally, only accurate tests in experimental apparatus simulating the real behaviour of the engine, both in terms of the materials present and the thermal processes, would be really indicative, but methods which adhere to such measures are very rare.

For this variety of reasons, many of the results available in literature often appear contradictory and only offer a general indication of the thermal stability of the fluid. Figure 3.2 provides values for the maximum temperature of thermal stability, reported by various authors, as a function of the standard binding energy. Whilst the standard binding energy of the atoms that compose the molecule is certainly not the only index controlling the temperature of thermal stability (as we mentioned, the type of materials present, for example, has a major importance), the diagram still gives us useful general indications: the fluids with an aromatic

Fig. 3.2 Approximate values of the limit temperature of thermal stability for various working fluids. The temperature values are taken from [12, 13, 18–22]

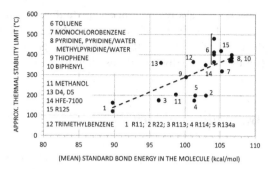

structure (toluene, pyridine, biphenyl) tend to have the highest thermal stability (about 400 °C); the linear hydrocarbons, under favourable conditions, seem to have limit temperatures of thermal stability of 300–350 °C; and, in general, the cyclical structure (in particular, the aromatic) of the molecule favours stability, more than an open structure. The presence of chlorine significantly reduces the thermal stability (for instance, for monochlorobenzene the maximum temperature of stability is around 300 °C, compared to 400 °C for toluene); by contrast, fluorine atoms, in place of hydrogen or chlorine, significantly raise the thermal stability: for example, refrigerant 125 (pentafluoroethane) and refrigerant 113 (1,1,2-trichloro-1,2,2-trifluoroethane) have limit temperatures of about 400 °C and 200 °C, respectively. The cyclic methylsiloxanes appear to have maximum operating temperatures close to 400 °C and the linear methylsiloxanes 50–100 °C lower. The alcohols seem to be usable up to maximum temperatures of 350–370 °C.

Health, Safety and Environmental Characteristics

A working fluid and its decomposition products ought not to be toxic, carcinogenic or particularly inflammable (or explosive). In fact, inflammability need not be a serious problem provided the appropriate precautions are taken and any potential ignition sources are removed from the critical zones of the plant (setting up the electrical system correctly and putting possible danger points in tubes—the flanges, for example—using fluid sensors and fluid collectors in the case of leakage). Auto ignition is, generally, a bigger problem and so is the risk of reaching explosive concentrations in the air.

From a strictly environmental point of view, the major worries are the ozone depletion potential (ODP), the global warming potential (GWP) and the atmospheric lifetime. For purely environmental reasons, then, numerous cooling fluids with excellent thermodynamic properties have been banned (for example, refrigerant 11, refrigerant 113 and refrigerant 114) and others are due to be phased out between 2020 and 2030. The so-called "natural" fluids, non-halogenated hydrocarbons, ammonia and carbon dioxide, favoured from a strictly environmental point of view, do not always present adequate thermodynamic characteristics. Among the hydrocarbons, in principle, it is relatively easy to find the most suitable fluid thermodynamically, but there remains the problem of their high inflammability and (for the aromatic hydrocarbons) their toxicity.

Availability and Cost

The fluid should be easy to purchase (maybe from more than one producer) and at a reasonable cost, because this will have a bearing not only on the start-up costs but also on the operating costs through possible leakages and the need for makeup during operation.

3.3 The Thermodynamic Aspects of the Organic Rankine Cycles

Figure 3.3 shows the typical plant layout for an engine with an organic fluid Rankine cycle. The working fluid, which is saturated, superheated or even in supercritical condition, begins its expansion in the turbine from point 4. At the outlet of the turbine, the steam temperature could still be relatively high and, before sending the vapour to the condenser, it could be worth cooling it (from 5 to 6) by means of a recuperative heat exchanger (a regenerator), in such a way as to preheat the liquid originating from the pump prior to sending it to the vapour generator (preheating from point 2 to point 3).

The regenerator is necessary for fluids with high molecular complexity (high σ parameter; see Sect. 2.5). In fact, the expansion of the working fluid, which is isentropic, if ideal, develops from point 4 (corresponding, for example, to saturated vapour) to point 5 in the zone of the superheated vapour (see Fig. 2.5), if σ is positive and, the higher σ is, the less the cooling of the fluid during expansion: assuming that the steam in expansion is similar to a perfect gas, in (2.29) the term C_P^0/R is equal to $\gamma/(\gamma - 1)$ and if γ diminishes, that is, σ increases, once an expansion ratio r_T is set, the gas will cool less (see Sect. 1.7.1 (1.24)).

The ratio C_P^0/R on which the σ parameter chiefly depends (see Fig. 2.6) is indirectly a function of the number of atoms N making up the molecule. From Fig. 2.6, we see that $\sigma \approx 0$ when $C_P^0/R \approx 10$ or, from Fig. 3.4, for a number of atoms $N \approx 5$–10. For $N > 10$, then, the upper limit curve of the fluid on the plane T–S generally has a positive slope and the expansions beginning with saturated vapour invariably take place in the zone of the superheated vapour.

In the condenser in Fig. 3.3, where the fluid passes from the thermodynamic conditions of point 6 to those of point 1, the vapour condensation is normally preceded by a section of de-superheating and in the vapour generator (the primary

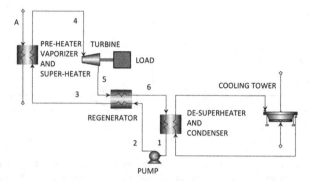

Fig. 3.3 Typical layout for an organic fluid Rankine engine. When the regenerator is not present, point 6 coincides with point 5 and point 3 coincides with point 2

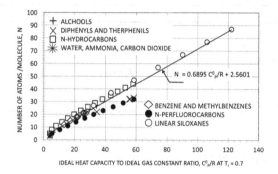

Fig. 3.4 Number of atoms that constitute the molecule, as a function of the ratio C_p^0/R for certain families of working fluids

heater), a significant percentage of the total heat exchange area is usually dedicated to the preheating phase of the fluid.

If the maximum pressure in the cycle is greater than the critical pressure P_{cr}, the phase change is not present and the fluid, gradually and steadily, passes from the liquid conditions 2 to the gas conditions 4 (crossing the Regions 6 and 7 of Fig. 2.1).

Generally, the heat $\dot{Q}_{in}$ for the cycle is process heat, available in the form of sensible heat and in Fig. 3.3 the fluid A identifies the heat source fluid.

As already observed (Sect. 2.5), of the numerous thermo-physical properties that characterise each fluid, in fact, only the critical temperature T_{cr} and the molecular complexity (quantifiable by means of the parameter of molecular complexity, σ) play a fundamental role in determining the characteristics of the thermodynamic cycle. Here below, we shall discuss their effects, strictly with regard to the thermodynamic aspects.

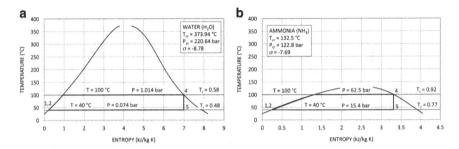

Fig. 3.5 Dependence of the characteristics of the thermodynamic cycle on its position along the limit curve. The two working fluids considered, water and ammonia, are both fluids with low molecular complexity ($\sigma < 0$)

Table 3.1 Some results for the two cycles represented in Fig. 3.5

	Water	Ammonia
Expansion pressure ratio	13.74	4.04
Expansion volume flow ratio	10.31	3.87
Expansion isentropic enthalpy drop (kJ/kg)	384.3	163.2
Turbine exhaust volume flow per unit power (m^3 s^{-1}/MW)	44.87	0.447
Evaporation heat/total heat	0.90	0.67
Preheating heat/total heat	0.10	0.33

The Effect of the Critical Temperature

The critical temperature T_{cr} of the fluid determines the position of the limit curve (and, therefore, of the conversion cycle which is created internally or, for the most part, within it) in the thermodynamic plane, and, having set the temperature levels of practical interest, it establishes in which region of the diagram of the fluid state the thermodynamic cycle will operate. By way of example, Fig. 3.5 reports the limit curves for water and ammonia and, within them, two conversion cycles are traced between the temperatures of 40 °C and 100 °C.

The water cycle uses a region of the state diagram far away from the critical point (at low reduced temperature $T_r = T/T_{cr}$) with low work pressures (prevalently, subatmospheric) and with the introduction of substantially isothermal heat. The ammonia cycle occupies a region of high reduced temperature (close to the critical point) with very high evaporation and condensation pressures and the introduction of heat which is largely not isothermal: see Fig. 3.5 and Table 3.1. The low value of the turbine exhaust volume flow per unit power of the ammonia (due mainly to the high condensation pressure) makes the cycle suitable for high-power engines.

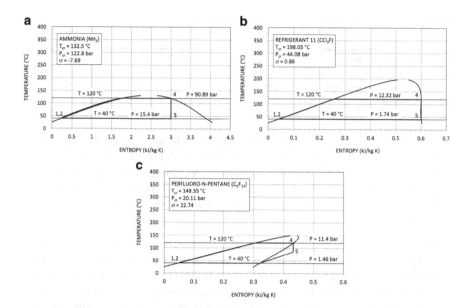

Fig. 3.6 Thermodynamic characteristics of the cycles as a function of the molecular complexity of the working fluid. The three fluids considered have similar critical temperatures, but very different parameters of molecular complexity

The Effect of the Molecular Complexity

The parameter of molecular complexity, σ, determines the shape of the limit curve and, therefore, that of the thermodynamic cycle which is inserted within it. As an example, Fig. 3.6 reports three thermodynamic cycles between the same extreme temperatures (120 °C and 40 °C) but which use three different fluids: the first, ammonia, with a simple molecular structure ($\sigma = -7.69$); the second, refrigerant 11, with a slightly more complex molecule ($\sigma = 0.86$); and the third, n-perfluoro-pentane, with a decidedly more complex molecule ($\sigma = 22.74$). It can be noted how, after an isentropic expansion, the first case produces a thick condensation; in the second, the condensation is negligible; in the third, there is a progressive superheating of the vapour as expansion occurs.

The consequences of such thermodynamic behaviour is important for the following reasons:

- For those fluids which do not give rise to any significant condensation during expansion in the turbine, there is no risk of erosion by the drops of liquid, which is a well-known danger in steam plants.
- The thermo-physical reason that keeps the risk of condensation at bay (the great molar heat, linked to the numerous degrees of freedom of complex molecule, that is, a high value of the C_P^0/R ratio; see (2.29)) also tends to make the specific heat of the liquid high and, therefore, its enthalpy, too, with the result that the

Table 3.2 Some results for the thermodynamic cycles of Fig. 3.6

	Ammonia	Refrigerant 11	Perfluoro-*n*-pentane
Expansion pressure ratio	5.88	7.08	7.80
Expansion volume flow ratio	6.13	6.79	10.51
Expansion isentropic enthalpy drop (kJ/kg)	182.25	36.86	18.77
Evaporation heat/total heat	0.46	0.64	0.35

heat is introduced to an appreciable degree also during the preheating phase of the liquid (compare Figs. 3.5a and 3.6a, c and see Table 3.2). This introduction of heat at lower levels than the maximum temperature, whilst penalising the thermodynamic cycle, also makes fractions of thermal energy at low temperature available for use, which, otherwise would have no use within the conversion cycle.

Table 3.2 shows some results for the thermodynamic cycles of Fig. 3.6. The expansion ratios are similar (the critical temperatures of the three fluids considered are, in fact, analogous: they differ only by around 15–20 %). The combined result of the great molecular complexity and the high value of the molecular mass is that the perfluoro-*n*-pentane cools down only slightly following expansion and provides a low specific work (18.77 kJ/kg, compared to a value of 182.2 kJ/kg for ammonia). This modest cooling during expansion of the perfluoro-*n*-pentane makes available an enormous quantity of heat that has been recovered from the regenerator (see Fig. 3.6c): the heat of the de-superheating is 44 % of the value of that of the condensation.

3.4 The Connections Between the Thermodynamics and the Machines

As we have seen and discussed in Sect. 1.6, the good performance of steam cycles is generally linked to a great complexity in the plant layout and in the prime mover (the turbine expander). Such complexity, while perfectly justified in the large thermoelectric power stations, cannot be transferred in practice to stations with more modest power levels (even at the level of a few dozen electrical MW there is a net simplification of the thermodynamic cycle and, often, a sharp drop in the quality of the expander). For power levels lower than several MW both technical and economic considerations suggest that the primary engine should have an internal performance that is not infrequently lower than 50 % and only exceptionally higher than 70 % (adopting multistage turbines with high revs—6,000–12,000 rpm). For power levels of several hundred kW it is not uncommon to have internal performances lower than 30–35 %.

The reason for such modest efficiency lies, firstly, in the high enthalpy drop that characterises the steam cycles (together with the technical and economic limitations linked to the peripheral speed and the number of stages) and, secondly, to the insufficient volumetric flows in the first phases of expansion.

Fig. 3.7 Example of the use of different regions of the thermodynamic plane (reduced) temperature–entropy for the organisation of thermodynamic cycles between the same temperature levels

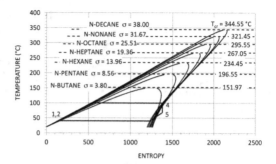

Having recourse to fluids other than steam (as already shown in the examples above) may bring about a significant reduction in the work of expansion, together with an adjustment of the volumetric flow to the minimum requirements of the turbo-expander. When the power level drops from the MW to just tens or even single kW, the possibility of choosing a fluid other than steam can make all the difference between feasibility and non-feasibility for a plant. The freedom to choose the levels of operating pressure (deriving from the freedom to choose the working fluid) makes it possible to obtain volume flows that are adequate to the turbomachinery.

To illustrate these aspects of interaction (thermodynamics—turbomachinery—heat exchangers), the following example may be useful [16]: consider a homogenous class of organic fluids (in our case, saturated hydrocarbons with a linear structure), whose components have a growing molecular complexity and, consequently, rising critical temperatures (see Fig. 3.7).

Use each of these fluids to make an engine of a given power (100 kW) between temperature levels that are also preset (maximum temperature $T_H = 100\,°C$, minimum temperature $T_C = 40\,°C$; see Fig. 3.7). The first result of such a choice would be the use of regions at different reduced temperature (T/T_{cr}) within the limit curve for the construction of thermodynamic cycles. Other direct consequences would be the reduction of the pressure levels (see (2.20), Fig. 2.3a and Sect. 2.5) and the temperature drop in the turbine with the increase in the molecular complexity (see Sect. 1.7.1 (1.24) and Sect. 3.3). The reduced cooling in the turbine means a growing benefit in the recovery of residual sensible heat in the vapour at the end of expansion by means of a regenerator.

The attendant effects on the thermodynamics of the cycle, the machinery and the regenerator are illustrated in Fig. 3.8a, b, and may be summarised as follows:

- The cycle efficiency tends to grow, if only slightly, as the molecular complexity grows in that, at lower reduced temperatures, the ratio evaporation heat/specific heat of the liquid increases and the preheating phase of the liquid assumes a relatively minor importance. By contrast, it is increasingly important to foresee the regeneration of the residual sensible heat at the turbine outlet, which grows rapidly as the molecular complexity increases (Fig. 3.8b). Only the fluids with a relatively simple molecule (like *n*-butane) can tolerate non-regenerative cycles (Fig. 3.8a).

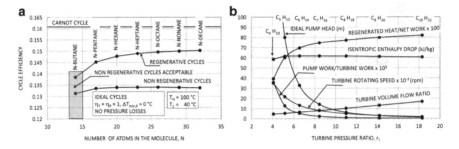

Fig. 3.8 Cycle efficiency and the various characteristics of the cycle and turbine as a function of the molecular complexity of the working fluid (in this figure, the molecular complexity is expressed by the growing number N of atoms, rather than by the factor σ of molecular complexity)

- The turbine pressure ratio grows continuously with the increasing molecular complexity, from 4 (*n*-butane) to about 18 (*n*-decane), causing an increase, albeit modest, in the work of expansion, despite the contrasting effect of the rise in molecular mass (see Fig. 3.8b). As the pressure ratio grows, so does the volume flow ratio (from about 5 to around 20).
- The angular speed of an optimised single-stage turbine decreases continuously as the molecular complexity grows, from 35,000 rpm for *n*-butane to 3,000 rpm for *n*-nonane, down to 1,800 rpm for *n*-decane (see Fig. 3.8b). Likewise, the average diameters of the rotor increase: from about 0.1 m to about 2 m.
- The work (and, therefore, the power) of the pump is fairly high for fluids with a simple molecule (around 4 % of that of the turbine in the case of *n*-butane) but rapidly becomes negligible as the molecular complexity increases. Coinciding with this (see Fig. 3.8b), for the final elements of the class considered (from *n*-octane to *n*-decane) it becomes feasible to eliminate the mechanical compression of the liquid, replacing it with direct feeding from the evaporator via the condenser, located higher up than the evaporator.

The reason for the great variation, as the fluid varies, in the number of revs (and the average diameters) of the turbine, made with just one stage for the sake of simplicity, is to be found in the greatly varying pressures of evaporation and condensation. The condensation pressure passes from 3.77 bar for the cycle with *n*-butane to 5.39 mbar for the cycle with *n*-decane, with a considerable increase in the volume flow at the outlet: from 0.226 to 86.5 m^3/s, respectively, for *n*-butane and for *n*-decane (for an isentropic power of 100 kW). Assuming a specific number of revs $\omega_S = 2\pi N/60 \times \sqrt{\dot{V}_{\text{out}}}/\Delta H^{3/4} \approx 0.43$ for the turbine stage at 3,000 rpm with *n*-nonane, with ΔH practically constant (see Fig. 3.8b) and one degree of reaction at the average diameter of 0.11, we get the notable variations in the number of revs and in the diameter mentioned above.

Very similar considerations to those made for the class of *n*-hydrocarbons can be applied to other classes of fluids (for example: chlorofluorocarbons, perfluorocarbons, aromatic hydrocarbons, siloxanes), and they all highlight the effectiveness of the choice of working fluid in controlling the characteristics of the thermodynamic power cycles.

Component Design: Turbines

As discussed in the example above (but also on the basis of what is illustrated in Sect. 1.7.2 with regard to closed ideal gas cycles), it is clear how the properties of the working fluid play an important role in the design of the turbine. In particular, the size of the enthalpy drop, the mass flow rate and the outlet to inlet volume flow ratio heavily influence the turbine characteristics.

The organic fluids, mostly with a high molar mass, are often characterised by modest values for the enthalpy drop (typically, 10–100 kJ/kg, compared to values of 500–1,500 kJ/kg in the traditional steam cycles), which leads to the possibility of using just one optimised turbine stage, or just a few stages, with a modest peripheral speed and low centrifugal stresses.

According to the cycle characteristics, the volume flow rate ratio may be highly variable: from just a few units for cycles at low temperature or positioned near to the critical point (even supercritical, but with high condensation pressure) up to 1,000 for cycles at high temperature (with a great difference between the maximum temperature and the condensation temperature) and with fluids having high molecular complexity. A large outlet to inlet volume flow ratio over a single stage gives a high number of Mach relative to the rotor inlet and an excessive variation in the blade height. Unusual values in the volume flow ratio often necessitate the exclusion of conventional stages with reaction degrees of 0.5 and have a further consequence in the significant variation in the speed triangles, passing from the root to the tip of the rotor blades.

The real gas effects, especially on the nozzles of the first stage, if the expansion begins near the critical point, may also lead to unconventional geometries of the inter-blade channels.

Figure 3.9b, for example, shows the factor of compressibility Z (see Sect. 2.3) as the expansion ratio varies during an isentropic expansion, with regard to the toluene cycle in Fig. 3.9a. If the expansion begins from a point near the critical point (case with $T_4 = 350\,°C$), the compressibility factor Z at the start of the expansion is worth around 0.4 and reaches values of nearly 0.9 only for expansion ratios $r_T \approx 10$; if the temperature at the start of expansion increases, the behaviour of the fluid volume approximates that of the perfect gas at higher pressures (for example, if $T_4 = 400\,°C$: $Z = 0.9$ when $r_T \approx 6$ and $Z \approx 0.6$ at the start of expansion). Figure 3.9c shows the area of passage (for unit mass flow of toluene) in the case of isentropic expansion, starting from $T_4 = 350\,°C$ and $P_4 = 55$ bar. Note how the hypothesis of an ideal gas tends to overestimate significantly the areas, especially the throat area (the minimum value).

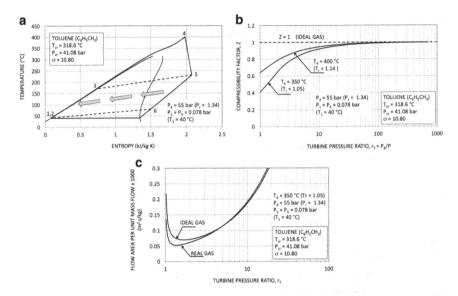

Fig. 3.9 (**a**) Supercritical cycle with toluene in the thermodynamic plane temperature–entropy. The plant layout is that shown in Fig. 3.3. (**b**) Behaviour of the compressibility factor during the isentropic expansion, as the expansion ratio varies. (**c**) Behaviour of the passage area during isentropic expansion for a unit mass flow of working fluid

In general, therefore, since the compressibility factor has a direct influence on the areas of passage for organic fluid cycles, the hypothesis of an ideal gas (with $Z = 1$) or, even more so, of a perfect gas (assuming, too, that C_P stays constant with the temperature; see Fig. 2.4) could prove to be misleading and lead to non-optimal solutions for the turbine.

Component Design: Heat Exchangers

The simplest evaporators to make are usually those of the pool-boiling type, which, thanks to their capacity for storing great volumes of fluid, are also less problematic in the power transitions. Should we need to minimise the fluid inventory (in the case of working fluids that are particularly expensive, for instance), the once-through solution is the most economical, even at subcritical pressures.

In condensers, the use of low integral fins is often a solution for reducing their size, which, otherwise, is penalised by the overall heat transfer coefficients which are modest for the organic vapours that condense (typically, in water condensers, 300–$1,200$ W/m^2 K for organic vapours compared to $1,500$–$4,000$ W/m^2 K for steam) and for the large sections of passage needed if the turbine discharges at low pressure.

Fig. 3.10 Diagram of the heat exchange for the regenerator (in ideal hypothesis) of the cycle in Fig. 3.9a. Working fluid: toluene

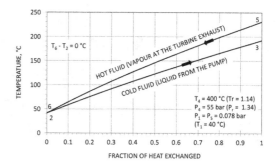

The regenerator is one of the peculiar components to the ORC. In those cycles with high ratios between the maximum and minimum temperatures and for fluids with high molecular complexity, the regenerator can dramatically raise cycle performance, which, however, will then be heavily dependent on the efficiency of the regenerator itself. The best configuration from the thermodynamic point of view would be that which is perfectly countercurrent, although this is difficult to achieve when the volume flows of the two fluids are very different. For the regenerators, just as for the condensers, the presence of a turbine at low revs with a high discharge area (low condensation pressures) may make the positioning difficult for given preset volumes.

Figure 3.10 shows the heat exchange diagram for the regenerator in the toluene cycle of Fig. 3.9a in the case where the minimum temperature difference $(T_6 - T_2)$ is null (ideal regenerator). Although the minimum temperature difference between the hot and cold fluids is equal to zero at one extremity (infinite heat exchange surface), the temperature T_3 does not reach the value T_5 anyway because the two fluids (the hot and the cold) have very different specific heat averages: about 1.56 kJ/kg K for the vapour at low pressure, originating from the turbine, and 1.96 kJ/kg K for the liquid originating from the feed pump. This difference in specific heats means that, even in the ideal case, the regenerator suffers from an intrinsic thermodynamic loss, which cannot be eliminated. That said, this will be amply outweighed by the overall thermodynamic benefit due to recovering (if only partially, in the real case) the heat which is ideally available $(H_5 - H_6)$ in Fig. 3.10 (which, in the case considered, is 1.16 times the useful work).

Figure 3.11 reports the performances calculated for saturated cycles, with and without regeneration, for several typical working fluids. The parameters assumed for the calculations are given in the figure. The results are indicative of the performances obtainable from ORCs at the maximum temperatures and condensation considered. Several thermo-physical properties of a thermodynamic interest for the fluids under consideration are collected in Table 3.3. As we can see, the working fluids with the high molecular complexity benefit substantially from the regeneration and the performances obtainable are, in this case, close to 27 %, with maximum temperatures of around 300 °C.

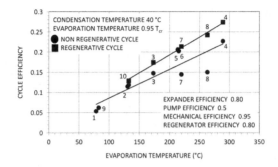

Fig. 3.11 Cycle efficiency as a function of the evaporation temperature for different working fluids and for cycle with and without regeneration. The numbers that identify the various fluids are the same as in Table 3.3

Table 3.3 Some thermo-physical data for possible working fluids in ORCs

No.	Fluid	Critical temperature (°C)	Critical pressure (bar)	Boiling temperature (°C)	Parameter of molecular complexity
1	Propane	96.68	42.48	−42.04	−1.24
2	n-Butane	151.97	37.96	−0.5	3.80
3	n-Pentane	196.55	33.7	36.07	8.56
4	Toluene	318.6	41.08	110.63	10.80
5	Methanol	239.35	80.84	64.7	−8.85
6	Ethanol	240.85	61.37	78.29	−5.09
7	MM[a]	245.55	19.14	100.52	30.75
8	MDM[b]	291.25	14.4	152.55	45.53
9	HFC 134a[c]	101.3	40.56	−26.07	−3.28
10	HFC 245fa[d]	154.05	36.4	15.3	6.147

[a]Hexamethyldisiloxane
[b]Octamethyltrisiloxane
[c]Hydrofluorocarbon (or refrigerant) 134a: 1,1,1,2-tetrafluoroethane
[d]Hydrofluorocarbon (or refrigerant) 245fa: 1,1,1,3,3-pentafluoropropane

   The performance of the cycle with MDM[24] (parameter of molecular complexity $\sigma = 45.53$), for instance, passes from a value of 15 % to 24 % in the case of a cycle with regenerator. Figure 3.12 represents the cycle on the thermodynamic plane temperature–entropy. As the graph makes clear, obtaining a good performance requires a massive regeneration (the heat regenerated is 2.5 times the useful work). Other peculiarities of the cycle are the relatively modest work of expansion (isentropic enthalpy drop is 100 kJ/kg) and the large volumetric expansion ratio (about 900, in the example considered). The modest turbine enthalpy drop is associated

---

[24]The octamethyltrisiloxane (MDM) is a member of the family of methylsiloxanes fluids, attractive working fluids for organic fluid cycles due to their technical characteristics: they are not toxic, only moderately inflammable and reasonably stable up to 300–350 °C (see [19]).

Fig. 3.12 Thermodynamic cycle with MDM in the thermodynamic plane T–S. The cycle performance is equal to 0.24. The parameters assumed for the calculation are those reported in Fig. 3.11

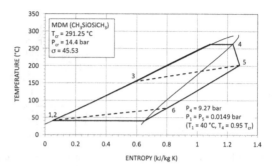

with the high volume flows at the outlet of the turbine. These characteristics are typical of all the working fluids with a molecular complexity similar to that of MDM.

Since most of the current applications use organic fluid Rankine engines principally for heat recovery, we shall discuss various aspects of the thermodynamics of the heat recovery in the next section.

3.5 The Heat Recovery: Basic Thermodynamic Considerations

Figure 3.13 represents a conceptual diagram of a system which consists of one heat recovery heat exchanger and one thermal engine. The engine uses the heat, available at temperature $T_{H,1}$, associated with a hot fluid with mass flow $\dot{m}_H$. In the drawing of Fig. 3.13, the engine uses a heat exchanger to cool the heat source fluid up to the temperature $T_{H,2}$ and converts the thermal power $\dot{m}_H (H_{H,1} - H_{H,2})$ into mechanical power $\dot{W}$. The engine (ideal) works between the temperature $T_H (\leq T_{H,2})$ and the temperature $T_C (\geq T_0)$. The thermal power $\dot{Q}_{out}$ is returned to the environment, at temperature T_0, by means of an appropriate heat exchanger.

The mechanical power $\dot{W}$ can be calculated by the exergy balance of Sect. B.1 (B.4), applied to the particular system considered here (in stationary conditions):

$$\dot{m}_H (H_{H,1} - T_0 S_{H,1}) - \dot{m}_H (H_{H,2} - T_0 S_{H,2}) - \dot{W} + \dot{E}_Q - T_0 \dot{S}_G = 0 \qquad (3.1)$$

In which, $\dot{E}_Q = -\dot{Q}_{out} (1 - T_0/T_C)$ (since the thermal power $\dot{Q}_{out}$ at the end of the process is released into the environment at the temperature $T_C > T_0$ assumed constant), $\dot{S}_G$ is the entropy generated in the unit of time in just the recovery heat exchanger (since the heat engine is presumed to be ideal) and $\dot{W}$ is counted with a negative sign since it is produced by the engine (exiting the system, not entering, as per usual in Appendix A).

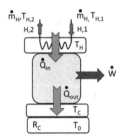

Fig. 3.13 Concept of a heat recovery engine. The heat source consists of fluid at the temperature $T_{H,1}$ which, in the figure, is cooled until the temperature $T_{H,2}$. The engine operates between the temperatures T_H and T_C. Environmental temperature is T_0

The term $\dot{S}_G$ is calculated from the entropic balance (B.3) applied to the heat recovery engine in stationary conditions

$$\dot{m}_H S_{H,1} - \dot{m}_H S_{H,2} + \dot{S} + \dot{S}_G = 0 \tag{3.2}$$

with

$$\dot{S} = (-1)\,\dot{m}_H \frac{H_{H,1} - H_{H,2}}{T_H}$$

From (3.2), we get

$$\dot{S}_G = \dot{m}_H \frac{H_{H,1} - H_{H,2}}{T_H} - \dot{m}_H \left(S_{H,1} - S_{H,2} \right) \tag{3.3}$$

The (3.3) substituted in (3.1) gives

$$\begin{aligned}
0 = {} & \dot{m}_H \left(H_{H,1} - H_{H,2} \right) - \dot{m}_H T_0 \left(S_{H,1} - S_{H,2} \right) \\
& - \dot{W} \frac{T_0}{T_C} \\
& - \dot{m}_H \left(H_{H,1} - H_{H,2} \right) \left(1 - \frac{T_0}{T_C} \right) \\
& - \dot{m}_H T_0 \frac{H_{H,1} - H_{H,2}}{T_H} + \dot{m}_H T_0 \left(S_{H,1} - S_{H,2} \right)
\end{aligned}$$

or

$$\dot{W} = \dot{m}_H \left(H_{H,1} - H_{H,2} \right) \left(1 - \frac{T_C}{T_H} \right) \tag{3.4}$$

The thermodynamic efficiency of the conversion of the thermal power $\dot{m}_H (H_{H,1} - H_{H,2})$ is therefore $\eta = (1 - T_C/T_H)$.

In general, the heat sources are always fluids that come from industrial processes and from general energy conversion processes (for example, products of the combustion at the outlet of a boiler and at the outlet of an internal combustion engine, gas from a blast furnace) and essentially, from the point of view of the heat transfer, the heat source behaves as (1) a fluid with finite (or equivalent) heat capacity rate $\dot{m}_H C_P$ (for example, liquids, gases or condensing vapours of non-azeotropic mixtures) or as (2) fluids with an infinite heat capacity rate, as, for example, the condensing vapours of pure fluids.

If a heat capacity C_P is definable for the fluid, the thermal power yield of the fluid can be written as $\dot{m}_H (H_{H,1} - H_{H,2}) = \dot{m}_H C_P (T_{H,1} - T_{H,2})$. Furthermore, $T_{H,2} = T_H + \Delta T_{min,H}$, with $\Delta T_{min,H}$, will represent the minimum difference in temperature (the "pinch point" temperature difference), in the heat recovery exchanger, between the hot fluid (the heat source) and the engine. Then

$$\dot{W} = \dot{m}_H C_P \left(T_{H,1} - T_{H,2} \right) \left(1 - \frac{T_C}{T_H} \right)$$

$$= \dot{m}_H C_P \left(T_{H,1} - T_H - \Delta T_{min,H} \right) \left(1 - \frac{T_C}{T_H} \right) \quad (3.5)$$

There exists, therefore, once the flow $\dot{m}_H$ has been set, a value of the temperature $T_H = T_{H,opt}$, corresponding to which the useful power $\dot{W} = \dot{W}_{opt}$ is the maximum:

$$T_{H,opt} = \sqrt{T_C (T_{H,1} - \Delta T_{min,H})} \quad (3.6)$$

On the other hand, should we wish to cool the heat source fluid down to the temperature T_0, there exists a limit value for the extractable power (equal to the variation in the physical exergy of the fluid, $\dot{W}_{optimal}$; see Sect. B.1) obtainable, for example, by calculating the performance from (1.13), assuming $\dot{S}_G = 0$:

$$\dot{W}_{optimal} = \dot{m}_H \left(H_{H,1} - H_{H,0} \right) \left(1 - T_0 \frac{S_{H,1} - S_{H,0}}{H_{H,1} - H_{H,0}} \right)$$

$$= \dot{m}_H C_P \left[(T_{H,1} - T_0) - T_0 \ln \frac{T_{H,1}}{T_0} \right] \quad (3.7)$$

Then, the ratio $\dot{W}_{opt}/\dot{W}_{optimal}$, having fixed $T_{H,1}$ and T_C and T_0, depends on $\Delta T_{min,H}$ and is invariably less than 1.0.

Figure 3.14 reports the ratio $\dot{W}_{opt}/\dot{W}_{optimal}$ as a function of the temperature difference $\Delta T_{min,H}$, for certain values of $T_{H,1}$. The ratio between the two powers decreases rapidly with $T_{H,1}$, having set a value of $\Delta T_{min,H}$. For example, for $\Delta T_{min,H} = 10\,°C$, the ratio between the pokers passes from 0.49, at $T_{H,1} = 400\,°C$,

Fig. 3.14 Ratio between the maximum power $\dot{W}_{opt}$ and the maximum power available $\dot{W}_{optimal}$, as the minimum difference in the heat recovery machine varies. The scheme referred to is that in Fig. 3.13

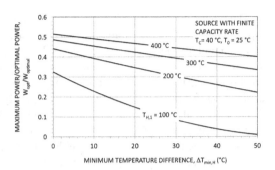

Fig. 3.15 Total thermal power passing through the engine per unit of useful power as the minimum difference in the heat recovery exchanger varies. The scheme referred to is that of Fig. 3.13

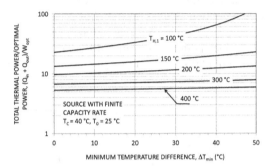

to 0.23 when $T_{H,1} = 100\,°C$. Even in the case with $\Delta T_{min,H}$ null, the maximum power does not equal the maximum power available: $\dot{W}_{opt}/\dot{W}_{optimal} = 0.32$ when $T_{H,1} = 100\,°C$ and 0.51 when $T_{H,1} = 400\,°C$. This is because the temperature $T_{H,2}$ is always greater than T_0.

The ratio $\dot{W}_{opt}/\dot{W}_{optimal}$ falls with the rise in $\Delta T_{min,H}$; the lower $T_{H,1}$ is, the faster this happens: when $\Delta T_{min,H}$ rises from 0.0 at 50 °C, the ratio passes from 0.32 to 0.01 when $T_{H,1} = 100\,°C$ (a fall of 97 %); it passes from 0.51 to 0.40 (a fall of 22 %) when $T_{H,1} = 400\,°C$.

Consequently, it is indispensable that the heat recovery exchanger be designed for a very small pinchpoint temperature difference, especially if the temperature $T_{H,1}$ is low. Otherwise, the ratio $\dot{W}_{opt}/\dot{W}_{optimal}$ will be excessively low.

In any case, the η_{opt} efficiency of thermodynamic conversion of the heat that is recovered will generally be fairly modest (even with $\Delta T_{min,H}$ null). As a result, the ratio

$$\frac{\dot{Q}_{in} + \dot{Q}_{out}}{\dot{W}_{opt}} = \frac{1}{\eta_{opt}} + \left(\frac{1}{\eta_{opt}} - 1\right) \tag{3.8}$$

tends to be high. In (3.8), $\dot{Q}_{in}$ and $\dot{Q}_{out}$ represent the thermal power transferred from the heat source to the engine and the thermal power released into the environment by the engine, respectively. The ratio $\left(\dot{Q}_{in} + \dot{Q}_{out}\right)/\dot{W}_{opt}$, which represents the total thermal power passing through the heat exchangers per unit of useful power produced, is shown in graph form in Fig. 3.15 as a function of $\Delta T_{min,H}$.

The ratio may reach values close to 100, for $\Delta T_{\min}$ excessively high, when $T_{H,1} = 100\ °C$; it stands at around 20 for $T_{H,1} = 100\ °C$ and $\Delta T_{\min,H}$ null. When $T_{H,1} = 400\ °C$, the ratio remains around values of 5–6.

Heat recovery plant are, therefore, characterised by the high cost of heat exchangers per unit of useful energy produced.

In plant engineering there will exist an optimal value of $\Delta T_{\min,H}$ that optimises the cost per unit of mechanical or electrical energy produced: for excessively low values of $\Delta T_{\min,H}$, the ratio $\left(\dot{Q}_{in} + \dot{Q}_{out}\right)/\dot{W}_{opt}$ does diminish, but the surfaces needed for the heat transfer (to guarantee a value which is finite) tend towards the infinite; for an excessively high $\Delta T_{\min,H}$, the η_{opt} conversion efficiency drops and the $\left(\dot{Q}_{in} + \dot{Q}_{out}\right)/\dot{W}_{opt}$ ratio rises and, with it, the cost of generating each unit of energy produced.

The $\dot{W}_{opt}/\dot{W}_{optimal}$ ratio is always lower than the unit on account of three thermodynamic irreversibilities (that is, three lost powers): the irreversibility at the level of the heat recovery exchanger, the irreversibility (the lost work) due to the missing reversible cooling of the source fluid down to environment temperature T_0 and the irreversibility caused by the temperature difference $\Delta T_{\min,C} = (T_C - T_0)$ (see Sect. 1.1 and Appendix C.2. As regards the choice of value to be assigned to $\Delta T_{\min,C}$, see Sect. 1.6.3):

$$\frac{\dot{W}_{opt}}{\dot{W}_{optimal}} = \frac{\dot{W}_{optimal} - \left(\Delta \dot{W}_1 + \Delta \dot{W}_2 + \Delta \dot{W}_3\right)}{\dot{W}_{optimal}}$$

$$= 1 - \frac{\Delta \dot{W}_1}{\dot{W}_{optimal}} - \frac{\Delta \dot{W}_2}{\dot{W}_{optimal}} - \frac{\Delta \dot{W}_3}{\dot{W}_{optimal}} \tag{3.9}$$

Where a specific heat C_P can be defined for the heat source fluid,

$$\Delta \dot{W}_1 = \dot{m}_H T_0 C_P \left(\frac{T_{H,1} - T_{H,2}}{T_H} - \ln \frac{T_{H,1}}{T_{H,2}}\right) \tag{3.10a}$$

$$\Delta \dot{W}_2 = \dot{m}_H T_0 C_P \left(\frac{T_{H,2} - T_0}{T_0} - \ln \frac{T_{H,2}}{T_0}\right) \tag{3.10b}$$

$$\Delta \dot{W}_3 = \dot{m}_H C_P \left(T_{H,1} - T_{H,2}\right) \frac{T_C}{T_H} \left(1 - \frac{T_0}{T_C}\right) \tag{3.10c}$$

with $T_{H,2} = T_H + \Delta T_{\min,H}$.

The maximum value of the ratio $\dot{W}_{opt}/\dot{W}_{optimal}$, or the maximum value of $\dot{W}_{opt}$ (having set $T_{H,1}$, $T_{H,2}$, T_C and fixed $\Delta T_{\min,H}$), is obtained when the sum of the three fractions of lost work (3.10a), (3.10b) and (3.10c) is least. In correspondence with this minimum, the temperature $T_{H,opt}$ will inevitably be equal to the value supplied by (3.6).

Fig. 3.16 Fractional work losses as a function of the maximum temperature T_H for a system with a recovery heat engine. The concept referred to is that of Fig. 3.13

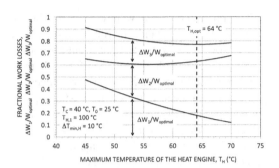

The ratios $\Delta \dot{W}_1/\dot{W}_{optimal}$, $\Delta \dot{W}_2/\dot{W}_{optimal}$ and $\Delta \dot{W}_3/\dot{W}_{optimal}$ are shown in graph form in Fig. 3.16 as a function of the temperature T_H, for $T_{H,1} = 100\,°C$ and $\Delta T_{min,H} = 10\,°C$. The value of $\Delta \dot{W}_1/\dot{W}_{optimal}$ is 0.48 at $T_H = 45\,°C$ and 0.12 at $T_H = 70\,°C$, as a consequence of the rapid increase in the η efficiency of the thermodynamic cycle of heat conversion. By contrast, $\Delta \dot{W}_2/\dot{W}_{optimal}$ increases by 0.17, for $T_H = 45\,°C$, till 0.56, for $T_H = 70\,°C$, as a consequence of the sudden increase in the thermodynamic loss, associated with the rising temperature $T_{H,2}$ of the source fluid. The sum of the three losses has a minimum that corresponds to the temperature $T_H = T_{H,opt} \approx 64\,°C$.

In general, then, we can state that the global conversion efficiency of the thermal power available (that which is supplied by the hot fluid, assuming that it can be cooled down to environmental temperature) comes from two equally important contributions: the thermodynamic quality of the heat engine's performance and the capacity of the heat engine itself (that is, of the working fluid that is used to make the thermodynamic cycle) to cool as much as possible the heat source fluid.

Strictly speaking, a fourth thermodynamic loss should be taken into account: the irreversible mixing of the heat source fluid with the environment at temperature T_0, at the pressure P_0 and with a fixed composition (the chemical exergy that is not used; see Sects. B and B.1). Generally, though, this final contribution is only modest and can be legitimately overlooked.

The Isothermal Source

If the heat source is at a constant temperature (as, for instance, in the case of condensing vapours of a pure fluid and assuming that we use only the latent heat made available), then $T_{H,1} = T_{H,2}$ and $\dot{m}_H (H_{H,1} - H_{H,2})$ represents the power yield of the heat source fluid during condensation. Therefore,

$$\dot{W} = \dot{m}_H (H_{H,1} - H_{H,2}) \left(1 - \frac{T_C}{T_H}\right)$$

$$= \dot{m}_H (H_{H,1} - H_{H,2}) \left(1 - \frac{T_C}{T_{H,1} - \Delta T_{min,H}}\right) \qquad (3.11)$$

Table 3.4 Some data regarding the fluids considered in Exercise 3.1

Fluid	Critical temperature (°C)	Critical pressure (bar)	Boiling temperature (°C)	Parameter of molecular complexity
Water	373.9	220.4	100	−8.777
CFC 11[a]	198.1	44.08	23.82	0.949
HFC 245fa[b]	154.1	36.4	15.3	6.147
CFC 12[c]	111.8	41.25	−29.79	−1.107

[a]Chlorofluorocarbon (or refrigerant) 11: trichlorofluoromethane
[b]Hydrofluorocarbon (or refrigerant) 245fa: 1,1,1,3,3-pentafluoropropane
[c]Chlorofluorocarbon (or refrigerant) 12: dichlorodifluoroethane

In this case, the maximum useful power and the maximum efficiency of the thermodynamic conversion cycle will be obtained when $\Delta T_{min,H} = 0.0$ (or $T_H = T_{H,1}$). Thus, if the heat source can be considered isothermal, from a thermodynamic point of view, those cycles are favoured in which the heat is used at the maximum temperature possible, namely, those where the evaporation heat of the working fluid is great compared to the preheating heat. This happens in two different circumstances (see Sect. 3.3): (1) when the thermodynamic cycle is located in the region of low reduced temperatures and (2) when, for given reduced temperatures of evaporation and condensation, a fluid with low molecular complexity is used for the cycle.

Exercises

3.1. With reference to a pressurised water source ($C_P = 4.184$ kJ/kg K) with $T_{H,1} = 150$ °C and for the fluids in Table 3.4, the $\dot{W}/\dot{W}_{optimal}$ ratio was calculated for variations in the evaporation pressure. The basic scheme of the thermodynamic cycle is that in Fig. 3.3, without a regenerator. The condensation temperature T_C is assumed to be 40 °C.

In Fig. 3.17, the ratio $\eta_{II} = \dot{W}/\dot{W}_{optimal}$ is represented as a function of the evaporation pressure. The η_{II} parameter can be interpreted as a second law efficiency, according to the definition (1.22), assuming the thermal power $\dot{Q}_{in}$ to be the thermal power available $\dot{m}_H C_P (T_{H,1} - T_0)$.

The evaporation pressure $P_{H,opt}$, which maximises the η_{II} ratio, rises from about 0.66 bar in the case of steam to 36 bar for HCFC 12. If the critical temperature of the working fluid is significantly higher than the temperature $T_{H,1}$, the optimal thermodynamic cycle will be located in the region of the low pressures and reduced temperatures (water and HCFC 11); as the critical temperature of the fluid approaches the value of $T_{H,1}$, the pressure and the reduced temperature of the optimal evaporation rise (HFC 245fa). If the temperature T_{cr} of the fluid is lower than the temperature $T_{H,1}$, the optimal cycle tends to become supercritical (the case of CFC 12).

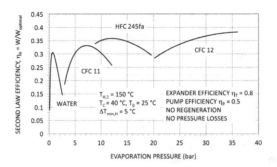

Fig. 3.17 Performance of the ratio between useful power and the maximum mechanical power obtainable from a heat source with variable temperature for different working fluids in thermodynamic cycles

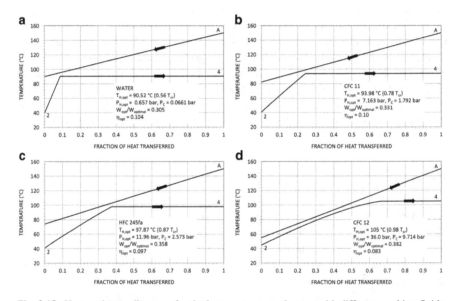

Fig. 3.18 Heat exchange diagrams for the heat recovery exchanger with different working fluids at the maximum power obtainable

The heat exchange diagrams in Fig. 3.18 clearly show how the greater value of the ratio $\eta_{II,opt} = \dot{W}_{opt}/\dot{W}_{optimal}$ is strictly correlated to the greater capacity of the working fluid for cooling the heat source. Although, for example, the steam cycle has an efficiency of 10.04 %, whilst the cycle with HCFC 12 has an efficiency of 8.3 %, the latter (cooling the heat source more) produces a useful power that is 25 % greater than the steam cycle.

Note how the operating conditions of the various cycles differ greatly. In particular, the steam cycle is completely subatmospheric.

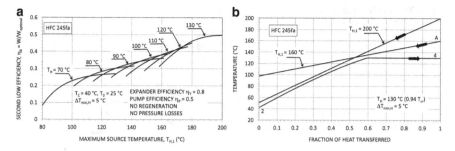

Fig. 3.19 (a) Efficiency of the second law for the thermodynamic cycle-heat source system, at varying maximum temperatures of the heat source $T_{H,1}$ and for different evaporation temperatures T_H for the working fluid (HFC 245fa). (b) Heat exchange diagrams for the thermodynamic cycle-heat source system

In general, from a strictly thermodynamic point of view, the cycle that exploits best a heat source of varying temperature (and that can be potentially cooled to nearly environmental temperatures) is the one which uses a working fluid of average molecular complexity, with a critical temperature close to the maximum temperature of the source.

3.2. Having selected a determinate thermodynamic cycle, the conversion efficiency of the heat introduced is fixed and so too is the law of heat introduction which, in the normal case of saturated cycles, has the trend shown in Fig. 3.18: a first part characterised by variable temperature (preheating of the liquid), followed by an isothermal part (evaporation of the liquid). Having fixed both T_H and T_C, the cycle can be used for various heat sources, each at a different maximum temperature $T_{H,1}$, characterised by different degrees of cooling and with distinct overall levels of energy quality (that is, different η_{II} efficiency).

Having fixed T_H and, usually, a $\Delta T_{min,H}$ in the recovery heat exchanger, the cooling of the source is unequivocally identified, as shown, for example, in Fig. 3.19b: the source at temperature $T_{H,1} = 160\,°C$ is cooled down to a temperature that is higher than that to which the source was cooled with temperature $T_{H,1} = 200\,°C$. Which of these thermodynamic cycle-source combinations will give the best result cannot be foreseen a priori, but must be determined, for different fluids and, for each of these fluids, for different temperature extremes of the cycle. One example of the results is given in Fig. 3.19a for the case of HFC 245fa.

Having set $T_H = 70\,°C$, for example, the heat source that is best exploited is that with temperature $T_{H,1} \approx 110\,°C$, with $\eta_{II} \approx 0.25$. The envelope of lines at the different evaporation temperatures of the working fluid gives best combinations between the cycles (saturated) and the heat sources for the specific working fluid under consideration (HFC 245fa). In the case we analysed, the maximum value of the second-law efficiency was obtained with a reduced evaporation temperature of around 0.94 ($T_H = 130\,°C$) and stands at about 0.5, for heat sources with $T_{H,1} \approx 190–200\,°C$.

Fig. 3.20 Simplified drawing
of the plant for Exercise 3.3

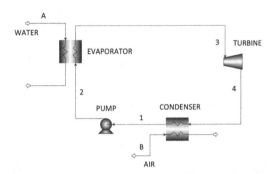

3.3. As discussed in Sect. 3.2, the choice of the most suitable working fluid for an organic fluid engine depends not only on its thermodynamic behaviour, but must also consider its thermal stability and its chemical compatibility with the materials used to make the plant.

In any case, the often inevitable decomposition should proceed relatively slowly, so that it does not substantially compromise the fluid characteristics and, thereby, deteriorate excessively the thermodynamic performance of the system. The effect of the working fluid decomposition on the thermodynamic efficiency of the engine is apparent firstly in a drop in useful power caused by the rise in condensation pressure, but the seriousness of the effects depends on the nature of the working fluid, its operating conditions and, clearly, the extent of the decomposition. Generally, when decomposition takes place, the products of the degradation have a simpler chemical structure than the initial working fluid, a lower molecular weight and single chemical bonds (for example, CH_4, CF_4).

With reference to a cycle with a hydrocarbon as working fluid (n-pentane, C_5H_{12}, critical temperature $T_{cr} = 196.55\,°C$ and critical pressure $P_{cr} = 33.7$ bar), we analyse the variations in the useful power of the engine when, starting from the design conditions, part of the fluid degrades and impurities of various kinds begin to form. The calculations refer to "off-design" situations, in the simplified hypothesis of stationary conditions. For the sake of simplicity, the performance of turbine and pump are also considered constant as the maximum and minimum operating pressures vary.

The case considered regards a hypothetical geothermal source of hot water at high pressure. The plant scheme for reference is that in Fig. 3.20. Figure 3.21 shows the performance of the useful power as the evaporation pressure varies.

The maximum value of useful power per unit flow of hot water is reached in correspondence with an evaporation pressure P_E of about 12 bar and is equal to around 80 kJ/kg of hot water. Starting from the design conditions ($P_E = 12$ bar, $P_C = 1.393$ bar, mass flow of hot water $\dot{m}_A = 0.8895$ kg/s (for unitary mass flow of working fluid), air flow to the condenser $\dot{m}_B = 32.14$ kg/s, $(UA)_E = 29{,}050$ W/K, $(UA)_C = 27{,}082$ W/K) and assuming that the global coefficient of heat exchange U_E at the evaporator and the global coefficient of heat exchange U_C at the condenser do not change, we get the results in Fig. 3.22a, b.

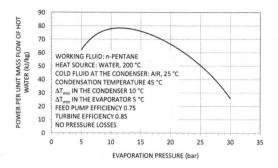

Fig. 3.21 Useful power as a function of the evaporation pressure

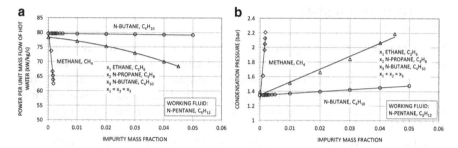

Fig. 3.22 Effect of the degree of impurity on (**a**) the useful power and on (**b**) the condensation pressure. The curves refer to the cases in which decomposition of the working fluid leads to the formation of just methane, just n-butane or a mixture of ethane, n-propane and n-butane

Figure 3.22a shows the eventual formation of very small quantities of methane are sufficient to compromise the plant operation: one mass fraction of methane of 0.002 reduces the useful power by 22 %, with a significant increase in the condensation pressure (from 1.393 to 2.2 bar; see Fig. 3.22b). By contrast, the formation of n-butane is less dramatic: one fraction of n-butane, equal to 0.05, reduces the useful power by less than 1 %, with a modest increase in the condensation pressure.

Where methane forms, though, its volatility makes it relatively easy to expel from the plant since it is particularly rich in the condenser, in the vapour phase (see Fig. 3.23). In this way, it is relatively simple to guarantee that the system operates properly. This would be completely different were the decomposition produces just n-butane: the vapour phase in the condenser is not particularly rich in it and its separation would, therefore, be rather difficult.

The conclusions in the example have a general qualitative validity, even though, the real effects of the decomposition of small quantities of fluid clearly depend on the type of fluid, the type of plant and the operating conditions (for example, super- or sub-atmospheric condensation pressures).

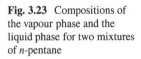

Fig. 3.23 Compositions of
the vapour phase and the
liquid phase for two mixtures
of *n*-pentane

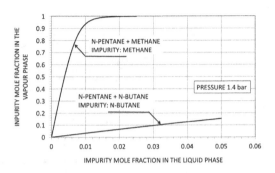

3.6   Some Examples of Applications of Organic Rankine Engines

There are numerous working fluids employed in Organic Rankine engines. They all have the potential to be thermodynamically adequate provided that they satisfy the requisites discussed in Sect. 3.2: adequate thermo-physical properties (thermodynamic and of heat exchange), appropriate thermal and chemical stability, compatibility with materials and substances present in the plant, safety of use and minimal environmental effects.

A list of the possible working fluids would not only not be exhaustive, but largely useless, given the number and range of families of organic products. In fact, every fluid should be considered and analysed specifically. For example, the toxicity depends greatly on the structure of the molecule and it can be misleading to generalise: the perfluorohexane (C_6F_{14}, of the family of perfluorocarbons), for instance, derived from hexane, substituting all the hydrogen atoms with fluorine atoms, used as a solvent and cooling fluid, is considered biologically inert and chemically very stable; the perfluoroisobutene (($CF_3)_2C_2F_2$, also called PFIB, a fluorocarbon alkene, product of the pyrolysis of the polytetrafluoroethylene) is more toxic than phosgene ($OCCl_2$, one of the most infamous chemical weapons used during World War I). Table 3.5 reports the parameters of health hazard, inflammability and chemical reactivity for certain fluid families and for several compounds that are typical of every family.

In fact, organic fluid engines mainly use the following working fluids: hydrocarbons, perfluorocarbons, siloxanes, refrigerants, fluoro-alcohols (for example, 2,2,2-trifluoroethanol). Potential working fluids could be all those fluids used as heat exchange fluids in the industrial sector: siloxanes, fluids with biphenyl or chlorinated biphenyl as base compounds, the perfluorocarbons or the perfluoroethers.

As far as the effects on the environment are concerned, these are regulated by a host of complex and varied mechanisms, and typical parameters include the atmospheric lifetime, the ODP and the GWP (see Table 3.6). In this case, too, each compound should be characterised separately, but, as a general rule, short atmospheric lifetimes give low ODP and GWP. The molecules that do not contain

Table 3.5 Some data about health hazard, inflammability and chemical reactivity for certain families of organic fluids

Fluid	Health hazard[a]	Flammability[a]	Chemical reactivity[a]
Hydrocarbons			
Alkanes, alkenes, alkyne, cycloalkane, alkadienes	1	4	0
Toluene	2	3	0
Naphthalene	2	2	0
Biphenyl	1	1	0
Siloxanes			
MM[b]	1	4	0
D5[c]	1	2	0
Perfluorocarbons			
Perfluorohexane	0	0	0 - 1
Perfluorodecalin ($C_{10}F_{18}$)	0	0	0 - 1
Refrigerants			
Ammonia	3	1	0
HFC 134a[d]	1	0	1
Diethyl ether ((C_2H_5)$_2$O)	2	4	1
Miscellanea			
2,2,2-Trifluoroethanol (CF_3CH_2OH)	2	3	1
Tetrachloroethylene[e]	2	0	0
Pyridine (C_5H_5N)	3	3	0
Chlorobenzene	3	3	0

[a]According to NFPA 704: a standard of the National Fire Protection Association that characterises the health hazard, the flammability and the reactivity on a scale from 0 (no hazard, normal substance) to 4 (severe risk)
[b]Hexamethyldisiloxane: $(CH_3)_3SiOSi(CH_3)_3$
[c]Decamethylcyclopentasiloxane
[d]Hydrofluorocarbon (or refrigerant) 134a: 1,1,1,2-tetrafluoroethane
[e]It is a liquid widely used for "dry cleaning" of clothes

iodine, bromine or chlorine usually have null ODP, but are often characterised by long atmospheric lifetimes, which could be a problem.

In this section, we shall present and briefly discuss several important examples and typical applications of organic fluid Rankine engines.

A Low Temperature Solar Plant

It is still not completely certain that low temperature solar thermodynamic systems (without concentration) can produce competitively priced electricity. In any case, for this to happen, certain minimum requisites need to be met: highly efficient solar collectors and conversion systems (the heat engine), modest consumption by auxiliary engines and favourable climatic conditions.

Table 3.6 Data regarding the environmental effects of certain fluids and compound families

Fluid	Atmospheric lifetime	ODP[a]	GWP[b]
Hydrocarbons			
Alkanes, alkenes, alkyne, cycloalkane, alkadienes	10–15 years	0	3–4
Toluene	2–500 days		
Naphthalene			
Biphenyl			
Siloxanes			
Volatile methylsiloxanes			
(linear and cyclic)	10–30 days	0	
Perfluorocarbons			
linear and cyclic	3,000–4,000 years	0	9,000–1,000
Refrigerants			
ammonia	5–10 days	0	0
HFC 134a[c]	15 years	0	1,450
Diethyl ether ($(C_2H_5)_2O$)	10 years	0	4
Miscellanea			
2,2,2-Trifluoroethanol (CF_3CH_2OH)			
Tetrachloroethylene[e]	5–6 months	0	0
Pyridine (C_5H_5N)			
Chlorobenzene			

[a]The ozone depletion potential (ODP) relative to refrigerant 11
[b]Global warming potential (GWP) for given time horizon of 100 years and relative to CO_2
[c]Hydrofluorocarbon (or refrigerant) 134a: 1,1,1,2-tetrafluoroethane

Anyway, the solar plant of Borj Cedria (Tunisia) [23] and Fig. 3.24, was built in the 1980s with the very intention of evaluating the technical and economic feasibility of systems of this type. The power plant was financed by the European Community also with the aim of transferring solar technology and training Tunisian engineers. The company Belgonucleaire of Brussels was assigned responsibility for the project, its construction and management. The solar collectors (flat, single glass with a copper absorbing sheet and selective coating) were completely built in Tunisia, whilst the engine was designed and built in Italy. The principal data for the plant, under design conditions, are reassumed in Table 3.7.

The working fluid chosen for the engine was perchloroethylene (or tetrachloroethylene, $Cl_2C=CCl_2$), a nonflammable fluid, with a high critical temperature ($T_{cr} = 346.85\,°C$), a critical pressure $P_{cr} = 44.9$ bar and a boiling point of 121.25 °C. Its molecular weight is 165.83 and the parameter of molecular complexity $\sigma = 4.0$. The perchloroethylene, due to its excellent solvent properties, is still widely used in the dry-cleaning industry and in the metal cleaning.

When, as in this case, the temperature difference between the hot source and the cold well ($T_H - T_C$) (that is, the difference between the evaporation and the condensation temperatures) is moderate, the thermodynamic performance does not represent the key parameter in choosing the working fluid, since any fluid with a

Fig. 3.24 Photograph of the ORC plant of Borj Cedria (1982) (from author [1])

Table 3.7 Principal data for the solar plant of Borj Cedria, under design conditions[a] [23]

Total surface of collectors	750 m^2
Temperature of entry/exit for the collectors' manifolds	85.4/100 °C
Temperature of entry/exit for the motor	98.5/86.5 °C
Cooling water temperature	20 °C
Cooling water mass flow	6 kg/s
Daily performance of the collector field	0.33
Electricity per day (net)	80 kWh
Volume of the storage hot water	45 m^3
Daily consumption of the auxiliaries of the field	13.6 kWh
Daily consumption of auxiliaries of the engine	1.6 kWh
Rated output power (net)	12 kW

[a]Daily insolation 5 kWh/m^2 day, corresponding to a reference clear day, next to the equinox

sufficiently high critical temperature will give an acceptable performance. As far as the final performance is concerned, the efficiency of the individual components is more important: the turbine, the heat exchangers, the feed pump, the speed reducer and the alternator. These components all depend on the nature of the working fluid and on the power size (see Sect. 3.4). In the case described, the working fluid is chosen for the very intent of optimising both the thermodynamics and, especially, the performance of single components.

Figure 3.25 shows the plant layout, taken here as reference. Given the modest molecular complexity of the working fluid ($\sigma = 4.0$), the regenerator is not necessary. The evaporation temperature T_4 under design conditions is about 84 °C and that of condensation T_1 is 30 °C. The gross electrical power is 16 kW; the net power (under nominal conditions) is 12 kW. From the calculations carried out according to the scheme in Fig. 3.25, assuming an efficiency of the expander of 0.84 and an efficiency of the reducer–alternator group of 0.833, we get a gross efficiency of 9.5 %. The reduced evaporation pressure (around 0.34 bar) is responsible for a

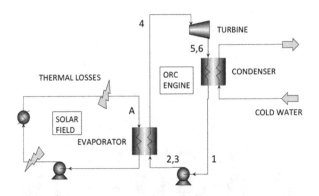

Fig. 3.25 Simplified scheme of the solar engine of Borj Cedria

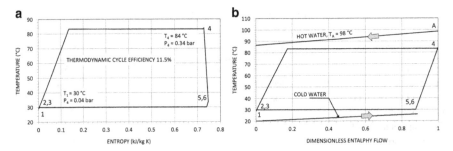

Fig. 3.26 Thermodynamic cycle of the solar engine of Borj Cedria. (**a**) Configuration of the thermodynamic cycle on the T–S plane. (**b**) Configuration of the cycle and heat exchange diagrams on the temperature–power plane

large volume flow at the turbine inlet and makes also it possible to eliminate the feed pump, substituted by the natural head of the liquid column, which feeds the evaporator, starting from the condenser located at an appropriate height (4–5 m).

The axial, mono-stage turbine, characterised by a large volume expansion ratio $\left(\dot{V}_{in}/\dot{V}_{out}\right)_S = 6.61$ and a modest specific work $(\Delta H)_S = 35.7$ kJ/kg, has a supersonic absolute vapour speed at the outlet of the fixed and rotating blades and subsonic relative speed at the rotor inlet and outlet. The rotation speed is 8,200 rpm. The maximum peripheral speed is about 180 m/s, with an average rotor diameter of 0.4 m and an average height of the rotor blade of about 1.8 cm.

In Fig. 3.26a the thermodynamic cycle is represented in the temperature–entropy plane and we see how the modest molecular complexity of the fluid and the modest difference between evaporation and condensation temperatures mean that the second-law efficiency of the cycle is very high ($\eta_{II} = 0.76$).

Figure 3.26b reveals the good match between the hot source and the cold source (the water for the condensation). The low maximum temperature of the heat source ($T_A = 98\,°C$) gives a modest efficiency of global conversion and the ratio $\left(\dot{Q}_{in} + \dot{Q}_{out}\right)/\dot{W}$ is 19.6. There are high costs for the evaporator and condenser, to which need to be added the costs of the solar field (also high, given the low efficiency of the collectors used).

The plant of Borj Cedria entered service at the end of June 1983 and operated correctly in both the solar and conversion sections. However, the cost of the energy generated proved to be very high, mainly due to the size of the heat exchangers and the extent of the solar field.

The Geothermal Binary Plants

The production of electricity from geothermal heat began thanks to the initiative of Piero Ginori Conti in 1904 at Larderello (Tuscany), with a geothermal steam engine and a dynamo of 10 kW. In 1913, the first commercial geothermal power plant was built (of 250 kW), again at Larderello, with the production and supply of electricity to the neighbouring towns.

In 1967, in the Soviet Union, one of the first binary power units was made at Paratunka, Kamchatka, with a heat source consisting of hot water at 80 °C, which, even today, represents one of the lowest temperatures ever exploited. Although the plant apparently functioned satisfactorily for many years, it was closed and dismantled towards the end of the 1970s—early 1980s due to leakage of the working fluid (the refrigerant R-12) [24]. Much earlier, in 1940, on the island of Ischia (Italy), a pilot binary geothermal plant was set up (see Sect. 3.1), but it never led to industrial development. An important example is the binary plant of 1.0 MWe at Nagqu, in North Tibet, at an average altitude of 4,500 m above sea level. It uses a geothermal source at 110 °C, with an air condenser, and became operative in 1993 [25, 26].

Today, about 11 % of the total 11,000 MW ca. installed power consists of binary plants: an organic fluid engine which, cooling the geothermal brine, produces electricity according to the usual scheme in Fig. 3.3, generally without a regenerator.

The binary systems are usually used when direct use of vapour is not possible and the temperature of the source, the geothermal brine, which is essentially liquid water with highly varying percentages of salts (NaCl, KCl, SiO_2), and gases (CO_2, H_2S) is not sufficient to make expansion of the geothermal fluid convenient (after the opportune flash). Shifting to binary cycle technology is considered convenient when the geothermal fluid consists mainly of water below the temperature of 150–180 °C. Looking to the future, if the forecasts and the technological endeavour now in progress for exploiting the engineered—or enhanced—geothermal systems (EGS) are correct and meet with success, then the interest in and potential of binary systems will certainly increase.

The binary system offers various advantages: it is generally more acceptable from an environmental point of view than any other geothermal power plant because the segregation of the geothermal fluid throughout the conversion process prevents the gas or other potentially polluting substances from being released into the environment; for sources with moderate temperatures (150–180 °C), they have better thermodynamic performances than the flash systems and can be used for the

Fig. 3.27 Energy quality of the conversion as a function of the source temperature, for various working fluids

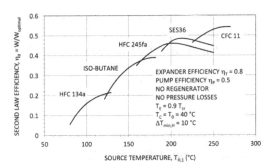

generation of electricity even with low temperature sources (90–100 °C); the binary systems reduce the problems connected with the scaling of fluids (the carbonate scale is prevented by installing downwell pumps, the silica scale is minimised, preventing the concentration growth caused by flashing). The cost of binary units is usually high but often compensated by the greater energy produced compared to flash systems.

The first European binary unit became operative in July 1992. The engine, with a gross electric power of 1.2 MW, was constructed by Italian companies under the EEC THERMIE programme and tested at Castelnuovo Val di Cecina, near to the geothermal power station of Larderello [27]. At the same time as the binary unit of Castelnuovo, there was that of "Travale 21" (in the district of Radicondoli, Italy): a binary power plant of 700 kWe.

In a binary plant, the choice of working fluid is a fundamental part of the project. Putting aside the characteristics of the heat exchange and the relative costs of the heat exchangers, at least two operative approaches are possible for the choice of working fluid: a fluid that will guarantee the highest conversion efficiency and a fluid that, once the power of the plant has been set, will permit the best design of the turbine. As we have already observed (see Sect. 3.5), if the heat source can be considered for the most part isothermal, optimisation of the thermodynamics is invariably favoured by a fluid with high critical temperature and a simple molecular structure. By contrast, when the heat source has a variable temperature, the maximum power is a combination of the engine fluid capacity to cool the source and, at the same time, the capacity of the thermodynamic cycle to best use the heat extracted from the source [28].

Figure 3.27 shows the performance of the second-law efficiency as a function of the maximum temperature of the geothermal water source for binary cycles using different working fluids. In Table 3.8, there are various thermodynamic properties of the fluids considered.

Each working fluid, according to its own critical temperature, exploits as best it can, the heat sources with maximum temperatures between the appropriate ranges. For instance, among the working fluids considered in the figure, HFC-134a would be advantageous from a strictly thermodynamic point of view for the exploitation of heat sources up to 120–130 °C, while the isobutane gives good thermodynamic

Table 3.8 Some data regarding working fluids used for the calculations in Fig. 3.27

Fluid	Critical temperature (°C)	Critical pressure (bar)	Boiling temperature (°C)	Acentric factor	Parameter of molecular complexity
HFC 134a	101.03	40.56	−26.07	0.3514	0.068080026
Isobutane	136.45	36.4	−11.72	0.183521	2.881631257
HFC 245fa[a]	154.05	36.4	15.3	0.375486	4.314297134
SES36[b]	177.55	28.49	35.64	0.3514	10.6252618
CFC 11[c]	198.05	44.08	23.82	0.189365	1.422963902

[a]Hydrofluorocarbon (or refrigerant) 245fa: 1,1,1,3,3-pentafluoropropane
[b]Solkatherm SSE36: an azeotropic mixture of refrigerant HFC 365mfc and a perfluoropolyether [29]
[c]Chlorofluorocarbon (or refrigerant) 11: trichlorofluoromethane

performances for sources from 120–130 °C up to 160 °C and so on. The refrigerant CFC-11, considered here just as an example, with a very high critical temperature, offers high second-law efficiency only if matched with sources with high maximum temperatures. Obviously, the various choices have a heavy influence on the turbine design. For example, if a source with maximum temperature of 100 °C is used with HFC-134a, in the examples in the figure, the condensation pressure will be 10.0 bar; if a source at 200 °C is used with HFC245fa, the evaporation pressure will be 2.42 bar. It follows, therefore, that a turbine at fixed revs would supply much greater power in the first case that in the second.

In general, a design project that aims at designing a low-pressure ORC means, at least for power ranges in the MW range, a turbine with a low number of revs can be used. By contrast, the solution of a high pressure ORC, using working fluids that operate with pressures close to critical pressures, requires, once the power is established, high-speed turbines (and feed pumps that consume more power).

In the case of the binary unit at Castelnuovo val di Cecina, the working fluid (the refrigerant CFC-11) was chosen to favour the design of the turbine. The CFC-11 was the only non-inflammable low-pressure fluid (with high critical temperature), and, in that period, the chlorofluorocarbons had not yet been phased out for environmental reasons.

One significant example of a geothermal plant with an ORC engine is the combined heat and power (CHP) plant at Altheim (Upper Austria). A geothermal district heating system had been in use since 1990 and the project of associating an ORC engine to the district heating was proposed to the European Commission in 1996.

The ORC turbogenerator and the well head, which are located near to the city in a densely populated area, required the use of a silent engine and a non-inflammable working fluid. One of the first to be completely built in Europe has a nominal power of 1,000 kWe. The ORC unit is cooled with water originating from a canal and the geothermal water is reinjected at a temperature of 70 °C into the deep geothermal Malm-aquifer [30]. The geothermal fluid is available at 106 °C and the high temperature downhole pump is installed at a depth of 250 m. The two

Table 3.9 Principal data of the Altheim power plant [31]

Source inlet temperature	106 °C
Source discharge temperature	70 °C
Source mass flow	81.7 kg/s
Electric power	1,000 kW
Generator speed	1,500 rpm
Cooling water inlet temperature	10 °C
Cooling water outlet temperature	18 °C
Cooling water mass flow	340 kg/s

Fig. 3.28 An image of the geothermal ORC engine at Altheim (2000) (from author [32])

wells (one for production and the other for reinjection of the geothermal water) are about 2,300 and 2,170 m deep, respectively. The ORC engine was installed in spring–summer 2000 and the first tests began in September 2000. The working fluid chosen was perfluoro-n-pentane (perfluoro-n-pentane, C_5F_{12}).

In May 2005, perfluoro-n-pentane was completely substituted by a new fluid (known by its commercial name Solkatherm SSE36), with good heat exchange properties and low viscosity and characterised by excellent turbine fluid dynamics. Heat stability is ensured up to 225 °C. The basic thermodynamic properties of SSE36 are reported in [29]. Solkatherm SES 36, an azeotropic mixture that is 65 % hydrofluorocarbon, HFC265mfc (1,1,1,3,3-pentafluorobutane) and 35 % perfluoropolyether (Galdem HT55), was used for the first time as a working fluid in the plant at Altheim, to replace perfluoro-n-pentane. Data from the plant have been summarised in Table 3.9 and a photograph of the engine can be seen in Fig. 3.28.

The availability of the plant is very high, over 7,500 h/year, but priority is given to district heating and just the hot geothermal water not sent to the mains is available for producing electricity.

The choice, made during the engine design, to create a low pressure ORC (that is, opting for a working fluid with a condensation pressure not too dissimilar to the atmospheric pressure and an evaporation pressure far from the critical pressure) has led to a turbine of 1 MW, with just 1,500 rpm. Around half the total cost of the plant was due to the well for reinjecting the geothermal fluid.

The CHP Biomass Plants

By the term biomass, we mean all the material resulting from living organisms, excluding those which, although the result of biological processes, have remained trapped underground so long as to merit the term fossil. In current usage, the term biomass refers only to biomass of vegetable origin. The use of biomass for the production of heat and electricity, typically in CHP units, for instance, in the way described in Sect. 1.4, passes through thermochemical processes that make the energy content, of a chemical nature, of the biomass available. Generally speaking, there are three thermochemical processes of practical application for this transformation:

- Combustion, where (in the presence of a significant excess of oxygen) bringing the biomass to a high temperature (1,000 °C), the physical and chemical structures of the biomass itself are demolished. The completely oxidised products of the combustion are, essentially, carbon dioxide and water.
- Pyrolysis, during which the biomass structure is broken down thermally into simple compounds, without the addition of oxygen, because the heating takes place without contact with the atmosphere. The final products are fuel gas (composed of hydrogen, carbon oxide, hydrocarbons and inert gases) and charcoal.[25]
- The oxidative gasification, which follows a similar method to combustion, but with an added oxygen in sub-stoichiometric amount, in such a way as to obtain final gas products that are not completely oxidised, which can then be used as fuel gas.

The fuel gases from pyrolysis or gasification can be used in ICE (Otto cycle), gas turbines or Stirling engines. The gases produced by the combustion process can be used for the steam production (to be used in steam turbines) or, typically via a secondary circuit with diathermal oil, constituting the hot source for an ORC turbogenerator. In principle, any prime mover capable of converting external heat could be used: even Stirling engines (see Sect. 1.8) and gas turbines (see Sect. 1.7).

---

[25]Charcoal, obtained by pyrolysis of vegetable biomass, was the only secondary fuel used by pre-industrial societies. Coke, produced by the pyrolysis of coal, was used in England during the 1640s and replaced charcoal in iron smelting towards the middle of the 1700s, when its production costs had dropped sufficiently to make it competitive.

In the case of Stirling engines, there is a need (in order to guarantee good performance) to keep the volumes destined to heat exchange small compared to the swept volume of the cylinder and this means using very compact heat exchange matrices, with small hydraulic radii and high surface to volume ratio. Since the combustion gases of biomass in a furnace are often heavy with particles particulate, sometimes, in order to prevent fouling of the hot heat exchanger, it is necessary to resort to systems of gasification rather than direct combustion of the biomass. In any case, the power of commercial Stirling engines is generally limited to several dozen kW.

Often, large quantities of virgin biomass are available as residual product from pruning, from virgin wood working, from periodic forest clearing and maintenance or directly from the cultivation of particular vegetable species. Wood is often available as waste from production processes (sawmills, furniture industry).

As we have mentioned, in the typical layout for a biomass plant using an ORC engine, the biomass is burnt in a furnace, using well-tested techniques, in a process that can be considered safe, reliable, clean and efficient. A heat transfer oil is used as the intermediate medium for the heat exchange between the combustion products (at 900–1,000 °C) and the working fluid for the ORC engine (which must necessarily have relatively modest maximum operating temperatures; see Sect. 3.2). Use of the heat transfer oil permits the furnace to function at low pressures, eliminating the need for licensed operators (on various shifts) to be present, as requested for steam systems in many European countries. The Organic Rankine prime engine converts the heat into mechanical energy. If the thermodynamic engine is designed appropriately, the condensation heat can be recovered in order to produce hot water (for example between 80 °C and 120 °C), for use in district heating and other uses, including industrial use (wood drying, sawdust drying) or for absorption chiller.

The special characteristics of the ORC technology (the simple start-up and switch-off procedures, the silent operation of the engine, the limited maintenance needed and the good performance with partial loads, the furnace characteristics) are such that, when the plant is operating in CHP mode, it is not necessary for specially qualified personnel, with a specific knowledge of the production sector and energy conversion, to be present [45]. The electrical power of ORC units for commercial biomass are usually just one or two MW.

Figure 3.29 shows, on the temperature–power plane, the heat sources (gas products of the combustion and the heat transfer oil), the organic fluid cycle and the heating process of the condensation water for a typical configuration of an engine using biomass. In Fig. 3.29b, for the sake of clarity, the cooling line for the combustion products has been eliminated. The calculations were made with reference to a unit flow of gas products of combustion. The working fluid for the organic fluid cycle is MDM (octamethyltrisiloxane, see Table 3.3). The evaporation pressure was assumed to be 10 bar and that of condensation to be 0.15 bar. The isentropic efficiency of the turbine was set at 0.8, the mechanical efficiency at 0.96. The isentropic efficiency of the pumps was 0.75 and the mechanical efficiency was 0.95. For the sake of simplicity, the pressure losses were not taken into account. The pinch point temperature differences were assumed to be 25 °C in

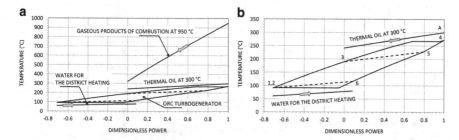

Fig. 3.29 Organic fluid engine for biomass. Configuration of the thermodynamic cycle and heat exchange diagrams in the temperature–power plane: (**a**) with the cooling line of the combustion products, which constitutes the heat source at high temperature, and (**b**) with just heat transfer oil, which decouples the organic fluid engine from the furnace. The working fluid for the engine is MDM (octamethyltrisiloxane)

the regenerator of the thermodynamic cycle and 15 °C for the condenser and for the secondary exchanger thermal oil-MDM.

The efficiency of the organic fluid cycle is 0.20 (considered as the ratio between the useful mechanical power and the thermal power made available by the diathermal oil). Ignoring the irreversibility of combustion (see Appendix B.1), the heat exchange exergy loss between gas products of the combustion and thermal oil is about 37 % of the exergy made available by the cooling up to $T_0 = 60\,°C$ of the combustion products. The exergy loss of the heat exchange oil-MDM is responsible for 0.7 % of the total losses, that of condensation for 7 %. The high exergy loss of the primary exchanger hot gas–thermal oil is due to the great difference in average logarithmic temperature between the two flows (279 °C). The thermal power regenerated per unit of useful mechanical power is about 3.5 (with a corresponding exergy loss of around 5.5 %).

Reducing the significant thermodynamic loss between gas products of the combustion and thermal oil is not easy and could, in principle, require the creation of systems with binary cycles (see, for example [33]), but careful consideration must be given to the increased plant costs that would be incurred in order to improve efficiencies.

One important example of an ORC biomass plant is the CHP plant of Lienz (East Tyrol, Austria) [34], which has been operative since 2001 and was built with financial funding from the European Commission within the fifth Framework Programme. The overall system consists of two biomass combustion plants: a hot water boiler, with nominal thermal power of 7,000 kW and a thermal oil boiler with nominal capacity of 6,000 kW.

The fuel used is biomass (wood chips, sawdust and bark, with a humidity content between 40 % and 55 %), originating from the nearby forests and the local wood industries. The electrical efficiency of the ORC unit (from 1,000 kWe, with hot water temperature at the condenser of 85 °C) is 18 %, which drops to 16–17 % and to 14 %, respectively, with a power of 50 % and 30 % of the nominal value: this testifies to the engine's great elasticity. Taking into consideration the plant

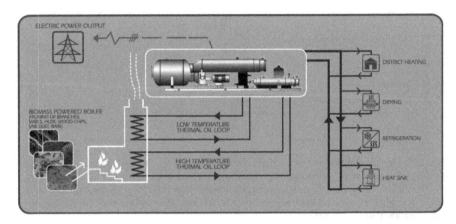

Fig. 3.30 Scheme for a biomass system with cogenerative organic fluid engine (with permission of Turboden srl)

Fig. 3.31 Photograph of the biomass plant with organic fluid Rankine engine at Fiera di Primiero in Trentino-Alto Adige (Italy). The engine installed (in cogenerative mode) has an electrical power of 1 MW (with permission of Turboden srl)

auxiliary systems, the (net) electrical efficiency drops to 15 % at the nominal power. In the winter season, the organic fluid thermal engine even works at 110–120 % of the nominal electrical load without prejudicing performance. At a nominal electrical load, the thermal power recovered at the condenser of the ORC unit is 4,440 kW. The biomass furnace is fitted with a preheater of the combustion air and an economiser for the thermal oil. The organic fluid engine works at full load for about 6,000 h/year.

Figure 3.30 represents a typical plant scheme for a CHP system with an organic fluid engine, Fig. 3.31 shows an ORC biomass plant in Trentino-Alto Adige (Italy) and Fig. 3.32 represents part of an organic fluid engine: the evaporator and the regenerator–condenser are shown.

Fig. 3.32 The biomass organic fluid Rankine engine at Varna, in Val d'Isarco (Italy). Cogenerative biomass engine with electrical power of 1 MW (with permission of Turboden srl)

Recovery of Waste Heat From Industrial Processes

The production cycles in many industrial sectors are often characterised by large quantities of waste thermal energy available in the process fluids. Typically gaseous fluids with highly variable mass flows and temperatures. After reuse of the fraction of thermal energy necessary for optimising the technological process, the remaining heat can be used to produce electricity (provided there are the necessary technical and economic conditions).

There are numerous sectors of potential interest: the cement industry (which typically provides heat from the combustion gases of the furnaces—downstream from the preheating of the raw materials—with temperatures of around 250–400 °C and from the cooling air for the "clinker", at temperatures lower than 300 °C), the steel-working industry (where the waste thermal energy is available from the process fumes and from fumes of the steel plants or foundries, often rich in powders), the glass industry (with availability of gases from the fusion of the glass at high temperatures of 400–600 °C), the petrochemical sector, ceramic production, heat recovery from the exhaust gases of diesel engines, etc.

In general, ORC technology enables heat recovery from any industrial process where waste thermal power is available in sufficient quantities to guarantee the design of engines with adequate efficiency at a reasonable cost.

With reference to strictly thermodynamic aspects, as illustrated in Sect. 3.5, in the case of a waste heat recovery system (WHES), with a heat source at variable temperature, the useful mechanical power per unit of available thermal power, the global efficiency η, is:

$$\eta = \frac{\dot{W}}{\dot{m}_H\left(H_{H,1} - H_{H,0}\right)} = \frac{\dot{W}}{\dot{m}_H\left(H_{H,1} - H_{H,2}\right)} \frac{H_{H,1} - H_{H,2}}{H_{H,1} - H_{H,0}} = \eta_C \times C_{TU} \quad (3.12)$$

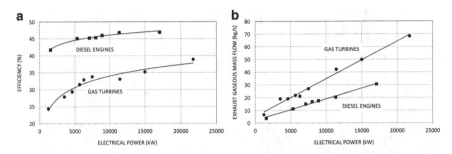

Fig. 3.33 A comparison between the characteristics of diesel engines and commercial gas turbine engines. (**a**) Efficiency as a function of electrical power. (**b**) Exhaust mass flow rate as a function of electrical power

with η_C representing the efficiency of the thermodynamic cycle and C_{TU} the coefficient of thermal utilisation. The global efficiency is, therefore, the result of two contrasting thermodynamic characteristics: the capacity to convert a high percentage of recovered heat into work (that is, a large η_C, which requires a high evaporation temperature T_H) together with a significant cooling of the heat source (which, inevitably, increases if T_H diminishes; see Sect. 3.5).

Having chosen the working fluid for the thermodynamic cycle, the consequences are that there is an optimal evaporation pressure and that, generally, the best working fluids have (1) a molecular complexity that is not too high and (2) critical maximum temperature close to the heat source (see Exercises 3.1 and 3.2). Ensured the thermal stability and taking into account the safety aspects of the fluid and its costs.

As an example of the specific application of ORC technology to heat recovery, we consider the case of recovering heat from the exhaust gases of a diesel engine.

The efficiencies as a function of the electrical power for some typical diesel engines and for gas turbines of similar power, in gen-set configurations, are reported in Fig. 3.33a. The electric efficiency of the diesel engines is always greater than the efficiency of the gas turbines of the same power: a difference of about 17 points per cent at 1.5 MW (about 25 % for the turbo gas and about 42 % for the diesel engine) and +11 points at 15 MW. The effect of the power size on the conversion efficiency is more pronounced in the gas turbines than in the diesel engines. For example, from 1.5 to 15 MW (a tenfold rise in power), the efficiencies of the diesel engines increase by a factor equal to 1.12 (from about 42 % to about 47 %) and the efficiencies of the gas turbines rise by a factor of 1.44 (from 25 % to 46 %). This is also a direct consequence of the difficulty in designing efficient, small and fast rotating turbomachines.

Figure 3.33b represents the mass flow of the exhaust gases as a function of the electric power for several diesel engines and gas turbines. At 15 MW, for example, the exhaust mass flow is about 27 kg/s for the diesel engine and 50 kg/s for a gas turbine of the same power: a value which is almost double. For the gas turbines considered in Fig. 3.33b, the mass flow rate per unit kW is 0.003 kg/s/kW: about 1.8 higher than that typical of diesel engines of similar power.

Fig. 3.34 Temperature and discharged thermal power for a typical heavy-duty diesel engine of 17 MW [35]

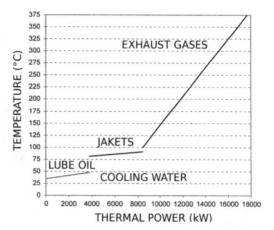

The smaller mass flows and the lower temperatures of the exhaust gases that are characteristic of the diesel engines (300–350 °C) compared with the gas turbines of similar power (400–500 °C) make the heat recovery more difficult.

Unlike gas turbines, in which the unconverted fraction of the chemical energy of the fuel is prevalently in the exhaust gases, in diesel engines it is discharged by (1) the exhaust gases (about 30 % of the LHV of the fuel), (2) the water cooling system of the jackets (and intercooler) (about 13 %) and (3) lubrication oil (about 10 %) [35].

Usually, (1) the exhaust gases have a maximum temperature of 350–400 °C and are at ambient pressure, (2) the water from the cooling of the jackets is available at about 90–100 °C, (3) the heat from the cooling of the lubrication oil is generally available at a temperature of about 50 °C: too low for an economic direct conversion into electrical energy. However, the heat at a lower temperature could be recovered in the heat exchangers of the plant to improve the overall energy balance.

With reference to a typical heavy-duty diesel engine of 17 MW, Fig. 3.34 reports the temperatures and the recoverable thermal powers from the exhaust gases and from the cooling water. The exhaust gases are available from $T_A = 375$ °C and are potentially cooled to 100 °C. The heat in the exhaust gases is about one half of the overall heat available and their relatively high temperature requires a recovery thermodynamic cycle operating with a working fluid of a high critical temperature.

In Fig. 3.35, we see a possible plant layout for recovering heat from a diesel engine. The working fluid of the Rankine cycle organic engine must have a high critical temperature and, as in the calculation example, toluene has been considered: $T_{cr} = 318.6$ °C, $P_{cr} = 41.08$ bar, $\sigma = 11.19$.

Several characteristics of the diesel engine used as reference are reported in Table 3.10. The composition assumed for the engine exhaust gases is the following (in molar fractions): 74.6 % nitrogen, 11.7 % oxygen, 6.7 % water, 5.9 % carbon dioxide and 1.1 % argon.

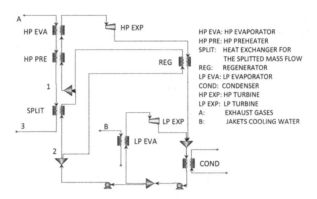

Fig. 3.35 Layout of a system for recovering heat from exhaust gases and from cooling water in a diesel engine [35]

Table 3.10 Several characteristics of the heavy-duty diesel engine considered for the calculations [35]

Electrical power	17,076 kW
Electrical efficiency	46.9 %
Temperature of the exhaust gases	374 °C
Exhaust gaseous mass flow	30.5 kg/s
Temperature of the jackets cooling water	80–91 °C
Cooling water mass flow	110.9 kg/s
Number of cylinders	18
Bore	460 mm
Engine speed	750 rpm

The cooling water in the engine, available at $T_B = 91$ °C, is cooled down to 81 °C on the low-pressure evaporator. The vapour of toluene (at 0.336 bar), expanding in the low-pressure turbine, produces 330 kW electric. The high pressure turbine (at 8.062 bar) generates 1,762 kW electric and the heat for preheating and the high pressure evaporation derives from the cooling of the exhaust gases of the engine from the temperature $T_A = 374$ °C down to $T_1 = 161$ °C, with direct heat exchange between the working fluid of the organic fluid engine and the gases of the diesel engine. In point 2, the flow of toluene is subdivided into two: one fraction (65 % of the flow on the high pressure turbine) goes over the regenerator and the remaining fraction, via the recovery exchanger denominated "split" in Fig. 3.35, further cools the exhaust gases of the diesel engine to the temperature $T_3 = 127$ °C. This introduces two thermodynamic advantages: a reduction in the irreversibilities of the heat exchange in the regenerator (see Fig. 3.10) and a greater cooling of the engine gases, which, otherwise, on account of the necessary presence of the regenerator, would be discharged into the environment at temperature T_1.

The net electrical power of the organic fluid engine is 2,072 kW, with a global efficiency of the combined system of diesel engine and organic fluid engine of 52.59 % (against a value of 46.9 % for just the diesel engine).

As far as the turbines are concerned, the high pressure one, with an isentropic enthalpy drop of about 170 kJ/kg, is rather loaded and, generally speaking, will require three axial stages; the low pressure one (with an isentropic work of about 44 kJ/kg) could be made with just a single axial stage.

Naturally, besides the purely thermodynamic analysis, a more detailed analysis of the return on investment would be needed to assess the economic viability of the plant (the cost of the heat exchangers, for example, is generally very high). However, the results of the above analysis give an idea of what can be expected from this kind of application, with a primary recovery of the exhaust gases and, possibly, the simultaneous recovery of cooling water from the engine.

In order to determine characteristics and costs of the ORC unit to be built, it is necessary to have accurate data from measurements of the temperature, the flows and about the nature and composition of the thermal sources. Apart from the nature of the fluid to which the thermal recovery is to be applied, the flows and temperatures must be noted in detail, including the variations they may undergo during the production process. The system available for the condensation (air or water) is also very important. In general, the size and costing of the heat recuperator exchanger require detailed data on the nature of the heat source (liquid or gas), such as its chemical composition, any polluting content (aggressive powders or chemical compounds) and the need to guarantee fixed temperature intervals (for instance, for the filtration or to prevent the appearance of incrustation and corrosion). On the basis of these characteristics, it is possible to establish the type of material needed, the exchange surfaces, the geometry and, consequently, the cost. Finally, it is important to determine the impact in terms of layout that the plant could have on the production process: modifying the existing lines could be especially costly, in particular, whenever there are problems linked to space availability and plant access.

As a general guideline, to be feasible, plant costs should not be greater than 3,000 euro/kW. The cost of electricity (difficult to predict in future years), the working time (in h/year) and expectations for the return on investment are all determining factors in the final choice.

The examples and the applications that we have presented and discussed (significative, although certainly not exhaustive) show how the choice of working fluid for each application is actually influenced by many factors: turbine enthalpy drops and flow rates, costs (of the fluid and the heat exchangers[26]), pressure levels, environmental compatibility, type of heat source, inflammability, type of cooling system, type of control system and supervision of the plant. In conclusion, the ideal fluid varies from case to case. Although, in principle, the organic fluid cycles adapt to every heat source (with maximum temperatures varying from 90–100 °C up to 300–400 °C) and to every level of power. Further research and development is, and always will be, needed to improve the heat exchangers, the expanders and in the study of new working fluids, possibly even multicomponent.

---

[26]The dimensions of the heat exchangers depend on the thermal power but also on the transport properties of the fluids, which directly influence the heat exchange coefficients.

3.7 Multicomponent Working Fluids for Organic Rankine Cycles

In principle, the working fluid used in an ORC could also be a mixture of two or more components: an azeotropic mix (for instance, the Solkatherm SES36; see Sect. 3.6) or a non-azeotropic mix. The non-azeotropic mixtures (see Exercises 2.2 and 2.1), once the pressure is set, are characterised by a difference, which may be more or less marked, depending on the fluids that compose the mixture and on the composition, between the dew and the bubble temperatures: the so-called temperature glide. That is, for the non-azeotropic mixtures, the liquid–vapour phase at constant pressure is not isothermal.

In practice, the presence of a temperature glide in evaporation could be useful from the thermodynamic point of view in those cases where the heat source is not intrinsically isothermal (as a possible alternative to the supercritical cycles), for example, in the heat recovery of flue gases exhausted by gas turbines or diesel and gas engines, or in the case of geothermal fluids and liquid-cooled solar collectors (see Sect. 3.5). If air is used for the condensation (with its modest heat capacity), the non-isothermal condensation may still be useful in reducing the flow of air necessary [36]. In the case of CHP systems for district heating, the condensation heat must be transferred to a liquid water loop, often characterised by a significant temperature difference between the delivery and the return from the heat consumers, and, in this case too, the use of heat engines with non-azeotropic mixtures could be thermodynamically appropriate.

By mixing several fluids, the critical point, too, varies continually (even in a non-linear way; see Exercise 2.2) and this characteristic may prove useful in creating thermodynamic cycles (see Sect. 4).

There are numerous fluid mixtures that could potentially be used (mixtures of hydrocarbons, fluorocarbons, polysiloxanes [36, 37] and, even, mixtures of non-miscible fluids [38]). Obviously, the criteria behind the choice of the best working fluid are still those briefly described in Sect. 3.2 and the aspects of interaction between the thermodynamics and the various components that make up the engine are still fundamentally those discussed in Sects. 3.3 and 3.4.

To highlight the peculiar characteristics of the non-azeotropic mixtures with a high temperature glide, in this section we consider, as a typical example, various cycles with ammonia–water mixtures of differing composition as the working fluid, assuming that the heat source is variable in temperature.

With reference to the scheme in Fig. 3.3 (a typical plant layout for an organic fluid engine), but without the regenerator, the performance of thermodynamic cycles has been calculated, considering ammonia–water mixtures of different molar composition as the working fluid. The heat source is water at 200 °C and the fluid used for condensation is air at 25 °C. For simplicity's sake, the water source is assumed coolable without any hindrance, down to room temperature. For the turbines and feed pumps, isentropic efficiencies of 0.75 are assumed and an efficiency of 0.95 to take into account the mechanical and electrical losses.

Fig. 3.36 Second law
efficiencies for cycles with
ammonia–water mixtures,
with different molar
compositions, as a function of
the evaporation pressure

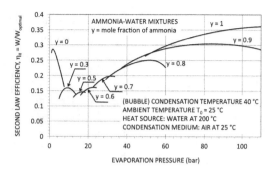

Again, for the sake of simplicity, pressure losses have not been taken into account. No limit was set a priori on the final temperature of the air necessary for the condensation. The logarithmic mean temperature differences on the exchangers have always been assumed, whenever it was possible to fix them beforehand, equal to 10 °C: a limit value that may be at the limit of technological practicability, but chosen in order to emphasise the importance of the heat exchangers, which, in the case of multicomponent fluids, are fundamental in order to highlight their thermodynamic characteristics.

Figure 3.36 shows the second-law efficiency η_{II} (for definition, see Exercise 1.2) for the various cases considered and for the cycles with pure fluids, water ($y = 0$) and ammonia ($y = 1$) as reference. The environment temperature T_0 has been taken as 25 °C for the purpose of calculations and the condensation temperature (that is, the condensation bubble temperature, in the case of mixtures) has been considered equal to 40 °C. The equation of state used to describe the thermodynamic properties of the mixtures is (2.5) with the mixing rules (2.30) and $\delta_{12} = -0.2589$.

The pure fluids (water and ammonia) tend to have higher η_{II} values than in the case of mixtures. This is a consequence of the different distribution of the entropic losses: on the machines, with regard to the heat source and the air used for the condensation and with regard to the environment taken as reference (at temperature T_0). The optimal η_{II} increases with the molar fraction y of ammonia and, at the same time, the optimal evaporation pressure rises.

Figure 3.37 shows just the losses in thermodynamic efficiency $\Delta\eta_{II}$ in the heat exchangers (heater and evaporator and condenser) at the maximum η_{II} for the mixtures considered. The mixture with molar composition equal to 0.7 in ammonia is that with the least entropic losses over the heat exchangers. The cycle with water vapour ($y = 0$) has, of all the cases considered, the highest loss in preheating and evaporation ($\Delta\eta_{II} = 0.035$) and the lowest loss in condensation ($\Delta\eta_{II} = 0.014$). For the ammonia–water mix with molar composition of 0.7, the losses over the heater and in condensation are 0.014 and 0.016, respectively.

The mixtures under consideration were shown to have an advantage in their single thermodynamic interaction with the sources. In particular, in the example discussed here, the mixture with $y = 0.7$ was the best. This has a direct consequence on the air flows requested by the condensation.

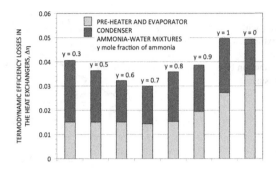

Fig. 3.37 Thermodynamic efficiency losses for heating and evaporation and for condensation in cycles with ammonia–water mixtures and with different molar compositions. Each case refers to the conditions of maximum second-law efficiency

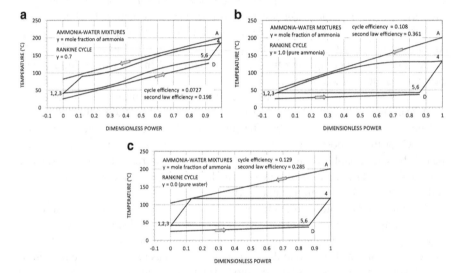

Fig. 3.38 Cycle configurations and heat exchange diagrams on the temperature–power plane, for different working fluids. The plant scheme for the cycle is that in Fig. 3.3, without regenerator. (**a**) Ammonia–water mixture. (**b**) Ammonia. (**c**) Water

Figure 3.38 represents, on the temperature–power plane, the thermodynamic cycles (each at the corresponding evaporation pressure of optimal second law efficiency) for the case of a working fluid consisting of an ammonia–water mixture, with molar fraction $y = 0.7$ in ammonia (Fig. 3.38a), for pure ammonia (Fig. 3.38b) and for water (Fig. 3.38c).

The heat exchange diagrams on the heater and the condenser justify the values of the various $\Delta\eta_{II}$ terms in Fig. 3.37. The air flows necessary for the condensation, per unit of gross useful power, are $0.119 \ \mathrm{kg \ s^{-1}/kW}$ for the case of the ammonia–water mix with $y = 0.7$, 0.688 kg s^{-1}/kW for pure ammonia and 0.565 kg s^{-1}/kW for water.

The air flow necessary for the condensation, gross useful power being constant, is, in the case of the steam cycle, around five times greater than that necessary for the condensation of the ammonia–water mixture; while the air flow necessary for the pure ammonia cycle is about six times greater. The power needed for pumping the air grows in direct proportion to the flow and, consequently, the net useful power diminishes. This could constitute enough of an advantage to justify the use of the mixture as the working fluid.

By contrast, though, the high temperature glide in condensation that is characteristic of the ammonia–water mixture in question is responsible, in the case of mixture with molar fraction $y = 0.7$, of the high final temperature of the air ($T_D = 127.3$ °C, in Fig. 3.38a). Above and beyond any problems this might give the plant, it leads to a significant thermodynamic loss (the exergy of the hot air is not to be considered a useful thermodynamic effect, unless the plant is cogenerative).

In the case of the three cycles, the condensation pressures are very different from each other: 0.0698 bar in the case of steam, 16.33 bar for pure ammonia and 10.32 bar for the ammonia–water mix with $y = 0.7$. The outcome of this is discussed in Sect. 3.4.

Appropriate modifications to the plant layout can raise the thermodynamic efficiency of the cycle in Fig 3.38a, with a mixture as the working fluid. Introducing a regenerator at the outlet of the turbine (according to the usual layout in Fig. 3.3) certainly increases the cycle efficiency and, in the case of the cycle in Fig. 3.38a, reduces the temperature T_D of the air at the condenser outlet. Introducing a regenerator also results in a minor cooling of the heat source, with a reduction of the useful power, and, ultimately, it is not sure that the second law efficiency will improve significantly. An alternative is to split the flow at the outlet of the feed pump, according to the scheme in Fig. 3.39a (and in Fig. 3.35, for a similar case of heat recovery at high temperature), seeking a good thermodynamic compromise between a sufficiently low temperature T_C and a reasonably high cycle efficiency (see Sect. 3.5).

Other configurations, partly repeating the plant layouts for absorption cooling cycles, foresee a separation of the steam phase, which is rich in ammonia, before the expansion, in such a way as to get a sufficiently low expansion temperature T_5. A typical example is the Maloney and Robertson cycle [39,40], of Fig. 3.39b (which also shows a superheater that is not present in any of the examples carried out here).

For example, in the Kalina cycle, in the version of Fig. 3.39c [40], which represents a more sophisticated version of the Maloney–Robinson cycle, the mixture flow available at the maximum pressure P_3 and at the evaporator composition $y_{1'}$ is further cooled by a second condenser, down to the temperature $T_{1'}$, before being sent to the heater and the boiler. Furthermore, there is a split in the flow after the low pressure compression.

The results of the second-law efficiency (for ammonia–water mixture with a base composition $y = 0.7$) are given in Fig. 3.39d. For comparison, the figure also shows

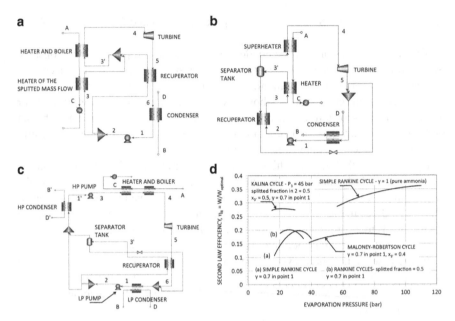

Fig. 3.39 Different plant configurations for cycles with ammonia–water mixtures. (**a**) Rankine cycle with flow split after the feed pump. A fraction of the high pressure flow passes across a regenerator (recuperator) and the rest cools the heat source. (**b**) Maloney and Robertson cycle. (**c**) A simple version of the Kalina cycle. (**d**) Results of second-law efficiency for cycles with ammonia–water as working fluid. In the case of the Kalina cycle, the abscissa represents pressure P_2. All the cases considered refer to saturated cycles, with a (bubble) condensation temperature of 40 °C, ambient temperature of 25 °C, heat source water at 200 °C and condensation air available at 25 °C

the curves relative to the Rankine cycle with $y = 1.0$ (pure ammonia). All the cases considered refer to saturated cycles. In the case of the Kalina cycle, the pressure in abscissa of Fig. 3.39d represents the pressure at point 2 in Fig. 3.39c.

Figure 3.40 gives the cycle configurations on the temperature-exchanged power plane, corresponding to the conditions of maximum second-law efficiency, with the corresponding values of the cycle efficiency and the second law.

Aside from the thermodynamic performance, when using multi-fluid mixtures, it should be noted that the coefficients of heat exchange are lower than those of the corresponding pure fluids [41–43]. In the specific case of ammonia–water mixtures, their use would normally be incompatible with carbon steels and aluminium for the problem of corrosion, so stainless steels or titanium are required instead [44].

The analysis above shows how the multi-component mixtures with a high glide temperature, among which the ammonia–water mixtures represent merely one example, are characterised by special thermodynamic properties, making it neither simple nor straightforward to choose which mixture to use or the plant set-up to adopt.

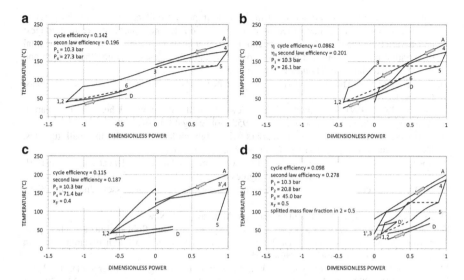

Fig. 3.40 Different configurations of the thermodynamic cycle on the temperature–power plane. (**a**) Rankine cycle with regenerator (plant scheme in Fig. 3.3). (**b**) Rankine cycle with regenerator and flow split after the feed pump (plant scheme in Fig. 3.39a). (**c**) Maloney and Robertson cycle (Fig. 3.39b). (**d**) A version of the Kalina cycle (Fig. 3.39c). All cases considered refer to saturated cycles, with a (bubble) condensation temperature of 40 °C, ambient temperature of 25 °C, heat source water at 200 °C and condensation air available at 25 °C

Certain plant configurations proposed are highly complex and all require sophisticated heat exchangers if they wish to make the best use of the specific thermodynamic properties of the non-azeotropic multi-component mixtures. Apart from the strictly thermodynamic considerations, the choice must also consider aspects directly linked to the plant and the machinery (operating pressures, fluid used for condensation, pump power).

References

1. Gaia M (2012) Thirty years of organic Rankine cycle development. In: First international seminar on ORC power systems, Delft TU-Technical University, The Netherlands, September 22–23 (Key-note presentation)
2. Rankine WJM (1859) A manual of the steam engine and other prime movers. Richard Griffin, London (Publishers to the University of Glasgow)
3. Anonymous (1827) Register of the arts and sciences – volume the fourth. A correct account of several hundred of the most important and interesting inventions, discoveries, and processes illustrated with about two hundred engravings. B. Steill, London
4. Stuard R (1829) Historical and descriptive anecdotes of steam-engines, and of their inventors and improvers, vol 1. Wightman and Cramp, London
5. Galloway E, Herbert L (1836) History and progress of the steam engine; with a practical investigation of its structure and application. To which is added, an extensive Appendix,

containing minute descriptions of all the various improved boilers; the constituent parts of steam engines; the machinery used in steam navigation; the new plans for steam carriages; and a variety of engines for the application of other motive powers, with an experimental dissertation on the nature and properties of steam and other elastic vapours; the strength and weight of materials, etc. Thomas Kelly, London

6. Ewing JA (1926) The steam-engine and other heat-engines, 4th edn. The University Press, Cambridge
7. Spencer LC (1989) A comprehensive review of small solar-powered heat engines: part I. A history of solar-powered devices up to 1950. Sol Energ 43(4):191–196
8. d'Amelio L (1935) The use of vapours with high molecular weight in small turbines. INAG – Industria Napoletana Arti Grafiche, Napoli (in Italian)
9. Spencer LC (1989) A comprehensive review of small solar-powered heat engines: part II. Research since 1950 – "conventional" engines up to 100 kW. Sol Energ 43(4):197–210
10. Duffie JA, Beckman WA (1991) Solar engineering of thermal processes, 2nd edn. Wiley, New York
11. El-Wakil MM (2002) Powerplant technology. McGraw-Hill, New York
12. Angelino G, Invernizzi C, Macchi E (1991) Organic working fluid optimization for space power cycles. In: Angelino G, De Luca L, Sirignano WA (eds) Modern research topics in aerospace propulsion. Springer, New York
13. Curran HM (1981) Use of organic working fluids in Rankine engines. J Energ 5(4):218–223
14. Gaudenzi P (1983) The low-boiling working fluids. Energie Alternative HTE 5(23):229–234 (in Italian)
15. Silvi C (2010) History of steam and electricity generation from solar heat by using flat or almost flat mirrors: research by Italian scientists since the 1800s. Energia Ambiente Innovazione 2:34–47 (in Italian)
16. Angelino G, Gaia M, Macchi E (1984) A review of Italian activity in the field of organic Rankine cycles. In: VDI Berichte 539 – Verein Deutscher Ingenieure. ORC-HP-technology. Working Fluid Problems. Proceedings of the international VFI-seminar, Zürich, 10–12 September 1984, pp 465–482
17. Various Authors (1958) Proceedings of the SRE-OMRE Forum, Held at Los Angeles, California, February 12 and 13, 1958. U.S. Atomic Energy Commission, Technical Information Service Extension, Oak Ridge (TID-7553. NAA-SR-2600)
18. Invernizzi CM (1990) Thermal stability investigation of organic working fluids: an experimental apparatus and some calibration results. La Termotecnica, pp 69–76 (in Italian)
19. Angelino G, Invernizzi CM (1993) Cyclic methylsiloxanes as working fluids for space power cycles. J Sol Energ Eng 115:130–137
20. Invernizzi CM, Pasini A (200) Thermodynamic performances of a new working fluid for power Rankine cycles. La Termotecnica, pp 87–92 (in Italian)
21. Angelino G, Invernizzi CM (2001) Real gas Brayton cycles for organic working fluids. Proc IME J Power Energ 215:27–38
22. Marciniak TJ, Krazinski JL, Bratis JC, Bushby HM, Buyco RH (1981) Comparison of Rankine-cycle power systems: effects of seven working fluids. ANL/CNSV-TM–87, DE82 005599
23. Gaia M, Angelino G, Macchi E, De Heering D, Fabry JP (1984) Experimental results of the organic fluid engine developed for the solar plant of Borj Cedria. Energie Alternative HTE 27(6):31–34 (in Italian)
24. DiPippo R (1979) Geothermal power plants of the Soviet Union – a technical survey of existing and planned installations. Contract EY-76-S-02-4051.A002, Southeastern Massachusetts University, North Dartmouth, MA and Brown University, Providence, RI
25. Cuellar G, Fangzhi Wu, Rosing D (1991) The Nagqu, Tibet, Binary Geothermal Power Plant, at 4500 m above sea level. In: Proceedings of the 13th New Zealand geothermal workshop, Auckland, 1991, pp 57–61
26. Schochet DN (2000) Case histories of small scale geothermal power plants. In: World geothermal congress, Kyushu, Tohoku, 28th May–10th June 2000, pp 2201–2204

27. Angelino G, Bini R, Bombarda P, Gaia M, Girardi P, Lucchi P, Macchi E, Rognoni M, Sabatelli F (1995) One MW binary cycle turbogenerator module made in Europe. In: Proceedings of world geothermal congress, Firenze (Italy), vol 3, pp 2125–2130, 18–31 May
28. Invernizzi C, Bombarda P (1997) Thermodynamic performance of selected HCFS for geothermal applications. Energy 22(9):887–895
29. Riva M, Felix Flohr, Fröba A (2006) New fluid for high temperature applications. In: Proceedings of international refrigeration and air conditioning conference at Purdue, 17–20 July 2006, paper R106, pp 1–8
30. Pernecker G, Uhlig S (2002) Low-enthalpy power generation with ORC-turbogenerator – The Altheim Project, Upper Austria. GHC Bulletin, March 2002, pp 26–30
31. Bombarda P, Gaia M (2006) Geothermal binary plants utilising an innovative non-flammable, azeotropic mixture as working fluid. In: Proceedings of 28th NZ geothermal workshop, November 15–17 2006, Auckland University, 6 pp
32. Gaia M (2006) Turboden ORC systems. In: Electricity generation from enhanced geothermal systems. Doc. 06A01720 - Turboden S.r.l., Strasbourg, 14 September 2006, 23 pp
33. Invernizzi CM, Paolo I, Sandrini R (2011) Biomass combined cycles based on externally fired gas turbines and organic Rankine expanders. Proc IME J Power Energ 215:27–38
34. Obernberger I, Thonhofer P, Reisenhofer E (2002) Description and evaluation of the new 1000 kWe organic Rankine cycle process integrated in the biomass CHP plant in Lienz, Austria. Euroheat Power 10:1–17
35. Pietra C (2008) Thermodynamic optimization of the energy recovery from diesel engines by means of organic Rankine cycles. Ph.D. thesis, Department of Mechanical and Industrial Engineering, University of Brescia (in Italian)
36. Angelino G, Colonna P (1998) Multicomponent working fluids for organic Rankine cycles. Energy 23(6):449–463
37. Chys M, van den Broek M, Vanslambrouck B, De Paepe M (2012) Potential of zeotropic mixtures as working fluids in organic Rankine cycles. Energy 44:623–632
38. Burnside BM (1976) The immiscible liquid binary Rankine cycle. J Mech Eng Sci 18(2):79–86
39. Maloney JD Jr, Robertson RC (1953) Thermodynamic study of ammonia-water heat power cycles. Oak Ridge National Laboratory, CF-53-8-43
40. Ibrahim OM, Klein SA (1996) Absorption power cycles. Energy 21(1):21–27
41. Arima H, Monde M, Mitsutake Y (2003) Heat transfer in pool boiling of ammonia/water mixture. Heat Mass Trans 39:535–543
42. Lu DC, Lee CC (1994) An analytical model of condensation heat transfer of nonazeotropic 1586 refrigerant mixtures in a horizontal tube. In: ASHRAE transactions: symposia OR-94-7-3, 1587 pp 5309–5318
43. Hong EC, Shin JY, Kim MS, Min K, Ro ST (2003) Prediction of forced convective boiling heat transfer coefficient of pure refrigerants and binary refrigerant mixtures inside a horizontal tube. KSME Int J 17(6):935–944
44. Zhang X, He M, Zhang Y (2012) A review of research on the Kalina cycle. Renew Sustain Energ Rev 16:5309–5318
45. Bini R, Manciana E (1996) Organic Rankine cycle turbogenerators for combined heat and power production from biomass. Presented at the third Munich discussion meeting "Energy Conversion from Biomass Fuels – Current Trends and Future Systems", Münich, 22–23 October 1996. Doc. 96A00412 – Turboden S.r.l., 8 pp

Chapter 4
The Real Gas Closed Cycles

In practice, all the mechanical energy produced by thermal engines is generated via cycles that use steam (see Sect. 1.6) or combustion products (in gas-turbines and internal combustion engines). One consequence of using these fluids is that thermodynamic cycles operate in well-defined regions of the diagram of state (see Sect. 2.1): either along the limit curve (in the case of steam cycles) or far from the limit curve, in the region of ideal gas (in the case of single-phase gas cycles).

The chance to use fluids other than steam or air in closed cycles brings into play two different lines of development: (1) the use of fluids in the same physical state as steam and air, but with particularly favourable properties (for example, chemical inertness at high temperature, great molar mass or low vapour pressure at turbine inlet conditions), and (2) the use of fluids in a different state of aggregation from that in conventional states (liquid, steam, perfect gas), localising the thermodynamic cycles in nonconventional regions of the thermodynamic plane, for example, in Regions 3, 4, 6, 7 and 9 of Fig. 2.1 (see Sect. 2.1).

Examples of the first line of development include the Rankine cycles, with organic fluid vapours (see Chap. 3), which, although they too are realised along the limit curve, exploiting the benefits of the phase change, use organic fluids in place of water; alternatively, the cycles with liquid metals, discussed in Chap. 5; or, further, the closed gas-turbine cycle (see Sect. 1.7) and the Stirling engine (see Sect. 1.8), which use different single-phase fluids in the region of the ideal gas.

In Fig. 4.1, on the temperature–entropy plane, several possible cycle configurations are drawn. The cycle on the limit curve (cycle A) represents the conventional Rankine cycle with steam as working fluid. Cycle B, an ideal gas cycle, represents the traditional Brayton closed cycle, with inert gas. Cycle C, a real gas cycle, is a closed cycle that works at minimum temperatures and pressures close to the critical point,[1] in such a way as to bring into play important effects of real gas

---

[1] For these types of cycle, point 1 at the start of compression may be positioned indifferently to the right of the upper limit curve (as in the case shown in the figure) or to the left of the lower limit curve.

C.M. Invernizzi, *Closed Power Cycles*, Lecture Notes in Energy 11, DOI 10.1007/978-1-4471-5140-1_4, © Springer-Verlag London 2013

Fig. 4.1 Condensation
thermodynamic cycles and
single-phase thermodynamic
cycles on the
temperature–entropy plane.
Each of the thermodynamic
cycles represented uses
different portions of the
thermodynamic plane with
different reduced
temperatures and pressures

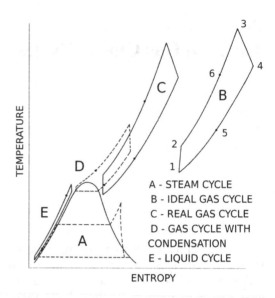

(discussed in Sects. 2.3 and 2.4). The gas and condensation cycle (cycle D) can be
defined as a liquid phase compression gas cycle, in that the waste heat rejection
takes place in the final zone of the saturation dome and the cycle extends almost
completely into the region of the supercritical gas. Finally, the liquid cycle (cycle E)
represents a thermodynamic cycle with a completely supercritical fluid in liquid
phase as working fluid.

The thermodynamic cycles shown in Fig. 4.1 are characterised by isobaric and
isentropic transformations, which represent the most common transformations for
continuous flow machines (see Sects. 1.3 and 1.5). The Stirling cycles are not shown
in the figure, but the same observations and conclusions apply to them, too (see
Sect. 4.3).

The position of the cycle in the different regions of the thermodynamic plane,
according to the various drawings in Fig. 4.1, once the thermodynamic conditions
(temperature and pressure) of point 1 have been set, is a direct consequence of the
values of temperature and critical pressure for the working fluid considered, that
is, the values of reduced temperature $T_{r,1} = T_1/T_{cr}$ and reduced pressure $P_{r,1} = P_1/P_{cr}$ of point 1.

Figure 4.2a, for example, shows in the thermodynamic plane temperature–
entropy, the saturation dome of three widely used fluids in industry, with similar
molecular complexity: air ($T_{cr} = -140.7\,°C$, $P_{cr} = 37.74\,bar$), carbon dioxide
($T_{cr} = 31.06\,°C$, $P_{cr} = 73.83\,bar$) and water ($T_{cr} = 373.95\,°C$, $P_{cr} = 220.64\,bar$).

By virtue of the Law of Corresponding States (see Sect. 2.3), the three limit
curves, rendered adimensional, converge into the single curve shown in Fig. 4.2b.
Having set, by way of example, a temperature $T_1 = 25\,°C$ and a pressure $P_1 = 100\,bar$, in the case of air, the corresponding point, in the reduced temperature–
reduced entropy plane, falls within the region of the ideal gas (see Sect. 2.1); with a
compressibility factor $Z \approx 0.97$, in the case of water, the point corresponds to the

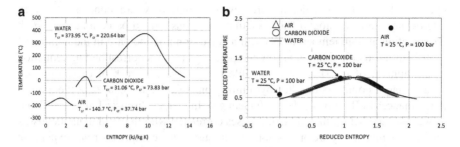

Fig. 4.2 (**a**) Limit curves for air, carbon dioxide and water in the temperature–entropy thermodynamic plane. (**b**) The same adimensional limit curves in accordance with the Law of Corresponding States

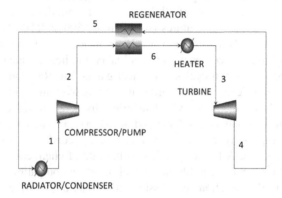

Fig. 4.3 Simplified scheme of the plant for a Brayton-type closed cycle. The scheme represents the thermodynamic cycles type C and E and the cycles type D of Fig. 4.1

thermodynamic conditions of a practically incompressible liquid and in the case of carbon dioxide the point falls very close to the critical point (see Fig. 4.2b) with a compressibility factor $Z \approx 0.25$, near the critical one. In the case of carbon dioxide, the thermodynamic conditions of the point at 25 °C and 100 bar are, therefore, characterised by the intense effects of the real gas, described in Sects. 2.2 and 2.4.

Therefore, each cycle shown in Fig. 4.1 can be realised, provided that the appropriate working fluid is chosen. In principle, then, specific thermodynamic cycles can be used for all the different applications that may be encountered.

The steam Rankine cycles are discussed in Sect. 1.6 and Chap. 3 is dedicated to the ORCs. The single-phase thermodynamic cycles with perfect gas are dealt with in Sect. 1.7 (the closed recuperated Brayton cycles) and in Sect. 1.8 (the Stirling cycles), whilst here we discuss the real gas closed Brayton cycles (cycle C of Fig. 4.1) and the liquid phase compression gas cycle (cycle D of Fig. 4.1). In fact, condensation has every right to be considered a real gas effect. Figure 4.3 represents the plant scheme to which we shall refer below.

A further effect of real gas, dissociation (Region 9 of Fig. 2.1), may also, in principle, be adequately exploited to create thermodynamic cycles [1, 2]. These cycles will not be considered here: even though they are conceptually very interesting, the technology of dissociating gases has still to be completely tested. Dissociation at high temperatures (with a corresponding increase in the number of moli) and the recombination at low temperatures (with a corresponding reduction in the number of moli) make it possible to increase the work of expansion and reduce that of compression, thereby limiting one of the main causes of the modest performances in Brayton cycles with ideal gas. In fact, as discussed in Sect. 1.7 and as we shall see below, the efficiency losses correlated with the fluid dynamic losses on the turbomachinery are proportional to the sum of the compression and expansion power, which, in the case of the Brayton cycles with ideal gas, is far greater than the useful work. The reactions of dissociation and association of the molecules comprising the gas mixture also brings into play the latent heats of reaction, which benefit the heat exchange and increase the coefficients of heat transfer in the heat exchangers. The result of all this is good thermodynamic efficiency even at not particularly high temperatures (for example 500–600 °C compared to the 800 °C typical of the Brayton closed cycles with ideal gas) and, having set the power, smaller and more compact turbomachinery and heat exchangers than those commonly used in Brayton cycles with ideal gas and in Rankine cycles. On the other hand, the chemically reactive gases for use as working fluids in Brayton or Rankine cycles, N_2O_4 and $NOCl$, which manifest dissociation and recombination in temperature intervals between ambient and 800–1,000 °C, are toxic. Compatibility with materials, especially at high temperatures, is a second serious problem, since the fluids are reactive. A further problem is the rate of chemical reaction and the times needed for reaching equilibrium, which must be compatible with the times corresponding to the variations in pressure and temperature in the various engine components.

It is worth remembering that, in the years 1970–1980, the then Nuclear Power Institute of the Bielorussian SSR Academy of Sciences developed a mobile nuclear reactor, the "Pamir-630D", of 630 kW electric with nitrogen tetraoxide as the heat-carrying fluid. The reactor, cooled by air, could be transported on a truck.

In the following sections of this chapter, we shall examine (1) energy cycles with carbon dioxide, as an example of closed cycles with real gas as working fluids: cycle C of Fig. 4.1 and cycle D of Fig. 4.1; (2) organic real gas cycles; and (3) the Stirling engines, the performances of which, in principle, would also benefit significantly from the real gas effects.

4.1 Carbon Dioxide Power Cycles

Carbon dioxide was used as coolant in the Calder Hall nuclear power station in 1956 [3]. Later, completely supercritical power cycles (single phase, with minimum pressure greater than the critical pressure) were discussed by Feher in 1967 [4]. Angelino, in a series of papers, [5–8] dealt exhaustively with the thermodynamics of supercritical cycles with carbon dioxide, even suggesting plant modifications to improve their performance. More recently, [9–12], for nuclear reactors, the

Fig. 4.4 Vapour pressure of
the carbon dioxide as a
function of temperature

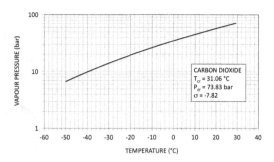

supercritical carbon dioxide cycles have been re-proposed as an alternative to
Brayton closed cycles with helium (see Sect. 1.7.7 and Exercise 1.8).

Carbon dioxide has a critical temperature of 31.06 °C, close to room temperature.
As a result, the Brayton thermodynamic cycles with temperature T_1 of 20–30 °C
and minimum pressures P_1 close to the critical pressure are characterised by
compressibility factors that are a third/quarter of the corresponding compressibility
factors for an ideal gas with the same minimum temperature and pressure. This
considerably reduces the work spent in compression and gives a useful work that it
still positive even though the maximum temperature of the cycle is not particularly
high.

The vapour pressure of the carbon dioxide is given in Fig. 4.4: the pressure of
the critical point is 73.83 bar. Therefore, the real gas Brayton cycles with carbon
dioxide tend to have high minimum and maximum pressures. On the one hand,
while the high density of the working fluid has beneficial effects on the size of the
machinery (see Sect. 1.7.6), on the other, it could increase the costs of the materials
employed. The internal efficiency of the turbomachinery tends to increase with
the pressure, but, having set the power and the specific work, the number of revs
also rises with the average operating pressure. As we shall see below, in the real
gas Brayton cycles, too, regeneration is hard, which makes them inappropriate for
conventional fossil power plants (due to the high stack heat losses they would incur).

The liquid phase compression gas cycle D in Fig. 4.1, having set the minimum
temperature T_1, has a minimum pressure lower than that of a totally supercritical
Brayton cycle (cycle B with pressure at the start of compression higher than the
critical pressure), with very comparable efficiency, and represents a thermodynamic
cycle with the advantage of the compression in the liquid phase (like Rankine cycles)
and with the expansion in the gas region, thereby avoiding the need for superheating
and re-superheating typical of the steam cycles (see Sect. 1.6). Compared to the
Rankine cycle with steam, then, the liquid phase compression gas cycle has far lower
expansion ratios and regeneration does not require the use of numerous exchangers
typical of the steam cycles.

Figure 4.5 shows the results of cycle efficiency for ideal and real cycles (the
performance of the ideal cycles are calculated with ideal components and in the
absence of pressure losses), as the compression ratio r_C varies. The maximum
temperature T_3 is always equal to 600 °C. The temperature T_1 has always been

Fig. 4.5 Thermodynamic efficiency for Brayton closed cycles with carbon dioxide. In all cases, it was assumed $T_1 = 22\,°C$ and $T_3 = 600\,°C$

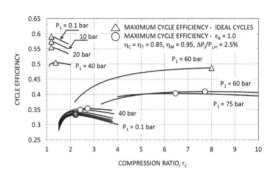

considered equal to that of condensation, at the pressure of 60 bar (about 22 °C, see Fig. 4.4). Thus, the cycle with P_1 equal to 60 bar represents a gas cycle with compression in the liquid phase, when P_1 is below 60 bar the Brayton cycles are to be considered in the zone of superheated gas and for pressures P_1 greater than the critical pressure, the Brayton cycles will be totally supercritical.

We chose a temperature T_1 that was lower than the critical temperature ($T_{r,1} = 0.97$) in order to analyse condensation cycles, too, and to emphasise the effects of the real gas in cycles with a single-phase fluid (Brayton cycles).

The same Fig. 4.5 also reports the cycle performance values for just the efficiency of the regenerator ϵ_R unit (for the definition of the regenerator efficiency, see Sect. 1.7.4, 1.35). In this case, the isentropic efficiencies η_C of compression and η_T of expansion are assumed to be 0.85 (the effect of reheating caused by fluid dynamic losses on the stages during expansion has been ignored for simplicity; see Sect. 1.7.3). Mechanical efficiency of the shaft has been considered as 0.95. The pressure loss on each exchanger has been considered equal to 2.5 % of the fluid pressure on entry into the component. There are two pressure losses on the regenerator: one on the low-pressure side and the other on the high-pressure side.

In the case of cycles containing just ideal components, it was observed that as the minimum pressure P_1 grows, the maximum efficiency of the cycle diminishes: the efficiency is 0.60 when $P_1 = 0.1$ bar for $r_C = 1.19$; it is 0.49 when $P_1 = 60$ bar for $r_C = 8$ (gas cycle with compression in the liquid phase).

The worsening of the thermodynamic quality of the ideal cycle as the pressure P_1 increases is the consequence of the negative effects of the real gas on the heat exchange in the regenerator. In fact, while the ratio between the net useful work and the total work $(W_T - W_C) / (W_T + W_C)$ increases significantly due to the positive effects of the real gas on the compressibility in the compression phase (see Fig. 4.6a), the notable difference in the specific heat at constant pressure between the high-pressure side and the lower-pressure side in the regenerator (see also Sect. 2.4, Fig. 2.4) introduces a temperature difference at the extremity greater than zero even when $\epsilon_r = 1$ (Fig. 4.6b).

Figure 4.6a shows the ratio between the net useful work $(W_T - W_C)$ and the total work $(W_T + W_C)$, under optimal performance conditions, passes from 0.46, in the case of an ideal gas, to 0.7 in the case of a liquid phase compression gas

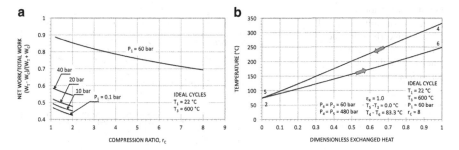

Fig. 4.6 Carbon dioxide ideal cycles, comparison between ideal gas and real gas (**a**) ratio between the useful work and the total work of turbomachinery (compressor and turbine); (**b**) diagram of heat exchange in the regenerator of an ideal cycle with compression in the liquid phase

cycle. Figure 4.6b reports the behaviour of the gas temperatures in the regenerator in the liquid phase compression gas cycle for the compression ratio $r_C = 8$. At the pressure of 60 bar, the average specific heat at constant pressure is 1,137 J/kg K, whilst, at the average pressure of 480 bar, the average specific heat is 1,677 J/kg K (48 % higher). The consequence is a temperature difference $(T_4 - T_6)$ which, rather than null, even with $\epsilon_R = 1$, stands at 83 °C. This introduces a thermodynamic loss which, in the case of ideal cycles where it is the only one present, penalises the cycle performance in the case of real gas compared to ideal gas.

The picture changes dramatically if we introduce the machine inefficiencies and the pressure losses. The results of cycle performance are shown in Fig. 4.5, for $\epsilon_R = 1$. The maximum cycle efficiency passes from 0.33 in the case of ideal gas, with a corresponding compression ratio of around 2.2, to 0.41 for the liquid phase compression gas cycle ($P_1 = 60$ bar con $r_C = 7.7$) and to 0.40 for the supercritical real gas Brayton cycle ($P_1 = 75$ bar with $r_C = 6.5$). In this case, the notable drop in thermodynamic losses on the work of the turbomachinery (the consequence of the increase in the ratio $(W_T - W_C)/(W_T + W_C)$, passing from the ideal gas to the real gas) improves the thermodynamic performances of the cycles with real gas, despite the adverse effect of the losses in regeneration.

Figure 4.7 illustrates the effects of the efficiency ϵ_R of the regenerator on the cycle performance for Brayton cycles with perfect gas and for cycles with compression in the liquid phase at $P_1 = 60$ bar.

The liquid phase compression gas cycles (although the conclusions are qualitatively valid also for the cycles C in Fig. 4.1 with real gas), suffer much less from the fall in efficiency ϵ_R than the cycles with ideal gas. For example, in the case of ideal gas cycles, when the efficiency of the regenerator ϵ_R passes from the unit value to 0.9, the maximum efficiency of the cycle drops by 25 %; in the case of the liquid compression gas cycles, the drop in the maximum efficiency is 10 % (in the intervals considered in the compression ratio r_C).

There is a comparison between the efficiency of the real gas supercritical Brayton cycles (cycle of type C in Fig. 4.1) and the liquid phase compression gas cycles (cycles of type D in Fig. 4.1) in Fig. 4.8. The figure also compares

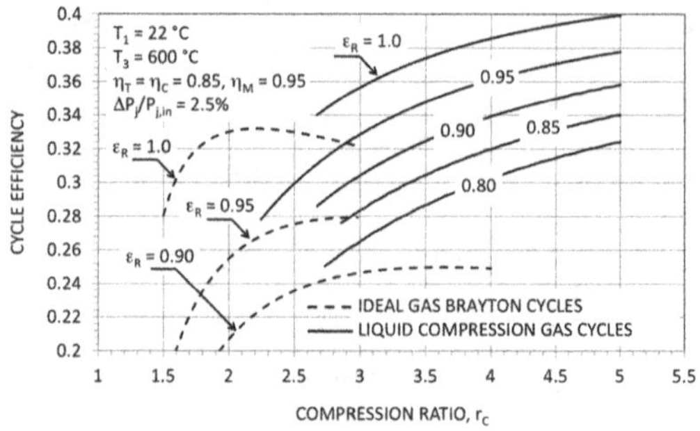

Fig. 4.7 Thermodynamic efficiency for Brayton closed cycles with carbon dioxide. Effect of the regenerator efficiency on the cycle performance

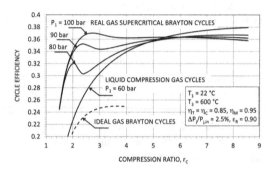

Fig. 4.8 Thermodynamic efficiency for Brayton closed cycles with carbon dioxide

the cycle performances with ideal gas. The maximum efficiency η for the liquid phase compression gas cycle is reached with very high compression ratios (in the case considered, the cycle efficiency eta reaches 38 % for $r_C = 8$); the totally supercritical Brayton cycles (with $P_1 = 80$, 90 and 100 bar) present a local maximum efficiency at low compression ratios (for instance, in the case with $P_1 = 90$ bar, $\eta = 0.35$ for $r_C = 2.3$) and a global maximum at high compression ratios (for instance, in the case with $P_1 = 90$ bar, $\eta = 0.36$ per $r_C = 6.0$).

This apparently anomalous thermodynamic behaviour is the direct consequence of the low temperature $T_{r,1}$ and is justifiable when we observe that, referring to the plant scheme in Fig. 4.3, the efficiency of the regenerator ϵ_R, in accordance with the definition given with (1.35), can be rewritten as

$$
\begin{aligned}
\epsilon_R &= \frac{H_6 - H_2}{H_4 - H_{5'}} \\[2mm]
&= \frac{H_4 - H_5}{H_4 - H_{5'}} \\[2mm]
&= \frac{H_4 - H_5}{(H_4 - H_5) + (H_5 - H_{5'})} \\[2mm]
&\approx \frac{\overline{C}_{P,4-5}\,(T_4 - T_5)}{\overline{C}_{P,4-5}\,(T_4 - T_5) + \overline{C}_{P,5-5'}\,(T_5 - T_{5'})}
\end{aligned}
\tag{4.1}
$$

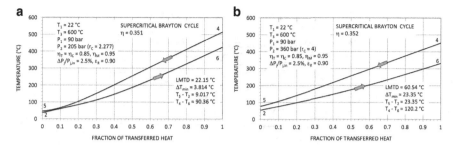

Fig. 4.9 Temperature trends in the regenerator of supercritical Brayton cycles with real gas. The pressure $P_1 = 90$ bar and the regenerator efficiency $\epsilon_R = 0.90$ are the same in both cases. (a) Cycle with modest compression ratio ($r_C = 2.28$). (b) Cycle with compression ratio $r_C = 4$. LMTD is the logarithmic mean temperature difference in the regenerator

with $\overline{C}_{P,4-5}$ the mean specific heat at constant pressure between point 4 and point 5 and $\overline{C}_{P,5-5'}$ the mean specific heat at constant pressure of the gas between point 5 and point 5'. Since $T_{5'} = T_2$, from (4.1) we get

$$T_5 - T_2 = (T_4 - T_5)\left(\frac{1}{\epsilon_R} - 1\right)\frac{\overline{C}_{P,4-5}}{\overline{C}_{P,5-5'}} \tag{4.2}$$

so, the temperature difference at the cold end of the regenerator, having set the efficiency ϵ_R, may also be a modest fraction of $(T_4 - T_5)$, according to the value of the ratio between the specific heats $\overline{C}_{P,4-5}/\overline{C}_{P,5-5'}$. The temperature difference $(T_5 - T_2)$ is then strictly correlated to the minimum temperature difference ΔT_{min} in the regenerator.

Just around the critical point, for $P_1 > P_{cr}$, when T_2, the temperature at the end of compression, is about 30–40 °C, the ratio $\overline{C}_{P,4-5}/\overline{C}_{P,5-5'}$ is significantly lower of the unit and $(T_5 - T_2)$ is small, with a ΔT_{min} that may turn out to be excessively low. See, for example, Fig. 4.9a.

Therefore, the heat exchange irreversibilities in the regenerator diminish rapidly with r_C due to the concomitant effect of a reduction in the Λ (logarithmic mean temperature difference) and a contemporary reduction in the power to regenerate per unit of useful work. As a consequence, the cycle efficiency grows rapidly as the compression ratio r_C increases.

However, with a rise in the compression ratio there also comes a rise in temperature T_2 and, from a certain point onwards, the ratio of the specific heats $\overline{C}_{P,4-5}/\overline{C}_{P,5-5'}$ starts rising again, to stabilise at more or less unitary levels. Consequently, the difference in temperature $(T_5 - T_2)$ grows, before diminishing slowly, stabilising at a more or less constant level. See, for example, Fig. 4.9b.

The thermodynamic irreversibility on the regenerator, anyway, continues to fall as r_C increases, but solely because the thermal power regenerated per unit of useful power is reduced. Figure 4.10, varying the compression ratio r_C, reports the thermodynamic losses (as a fraction of the maximum efficiency, see Exercise 1.2)

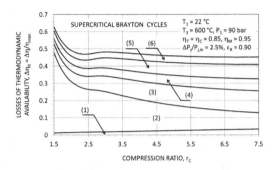

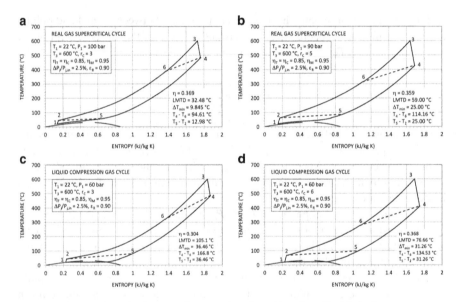

Fig. 4.10 Losses of thermodynamic availability as a function of the compression ratio r_C for supercritical Brayton cycles with carbon dioxide. (1) loss in compression, (2) loss on the regenerator, (3) loss associated with heat introduction, (4) loss in expansion, (5) loss on the radiator, (6) loss due to the mechanical efficiency of the power shaft

Fig. 4.11 Thermodynamic diagrams for several cycles with carbon dioxide in the temperature–entropy plane. (**a**) and (**b**) Supercritical Brayton cycles. (**c**) and (**d**) Cycles with compression in liquid phase

for a reference case. The figure clearly shows the diminution of the irreversibilities due to heat exchange in the regenerator: at first, very fast, then, after the inflexion (with r_C about 2.5), much more slowly.

Figure 4.11 reports two real gas supercritical Brayton cycles and two condensation cycles, on the T-S plane.

The supercritical Brayton cycles, in accordance with the results in Fig. 4.8, for relatively high pressures P_1, have good efficiency even at modest compression ratios

Fig. 4.12 Simplified plant
scheme for cycle with partial
condensation and split
compression

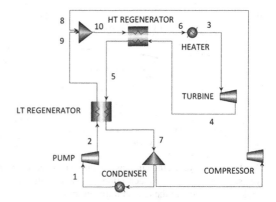

(for example if $P_1 = 100$ bar, when $r_C = 3$ there is an efficiency η of around 0.37).
In these conditions, though, the minimal temperature differences in the regenerator
(due to effects linked to the rapid change in specific heats at constant pressure,
discussed above) could be very small (about 10 °C, in the example in Fig. 4.11a). As
the compression ratio increases, the temperature profiles in the regenerator become
more regular; the efficiency is not too badly penalised by the high r_C, but the
maximum pressures are very high (in the example of Fig. 4.11b, $P_1 = 90$ bar, $P_2 =
450$ bar, $\eta = 0.36$ with $\Delta T_{min} = 25$ °C).

The condensation cycles, with compression in the liquid phase and expansion
starting from the supercritical gas zone, are characterised by good efficiency only
at high compression ratios (in the example of Fig. 4.11d, the efficiency $\eta = 0.37$
for a compression ratio $r_C = 6$). In these cycles, the high specific heat at
constant pressure of the compressed fluid compared to that of the gas exhausted
by the turbine, always introduces a high thermodynamic irreversibility into the heat
exchange at the level of the regenerator, in particular, at small compression ratios.
In the example of Fig. 4.11c, for $r_C = 3$, we get $\Lambda = 105$ °C, with $(T_4 - T_6) =
167$ °C and $(T_5 - T_2) = 36$ °C, with an efficiency η of around 0.3. This difference
in the specific heats between the regenerator fluids is responsible for the growing
temperature difference between the hot fluid and the cold fluid in the heat exchanger,
which, in the end, means heavily penalising the efficiency.

In the case of total condensation cycles, a substantial reduction in the irreversibilities in the regenerator is obtained by dividing the compression partly into a liquid
phase and partly into a gas phase [7], according to the plant scheme in Fig. 4.12.

A balance in the thermal capacities can be achieved in the LT regenerator by
dividing the low-pressure gas flow into two portions at point 7. If $\alpha = \dot{m}_1/\dot{m}_7$
represents the fraction of the flow that condenses, $1 - \alpha = \dot{m}_8/\dot{m}_7$ is the fraction
directly compressed from the pressure P_7 of point 7 to the pressure P_8 of point 8.
Having set the temperature differences at the ends $(T_7 - T_2)$ and $(T_5 - T_9)$, the
α fraction of the flow can be calculated by the balance of power on the low
temperature regenerator: $\alpha (H_9 - H_2) = (H_5 - H_7)$.

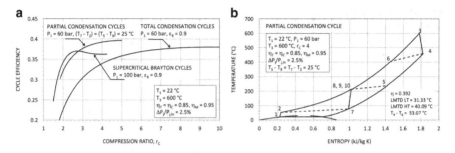

Fig. 4.13 (a) Cycle efficiency as a function of the compression ratio for supercritical Brayton cycles, cycles with total condensation and cycles with partial condensation. (b) Scheme of a partial condensation cycle in the temperature–entropy plane

The two flows, heated to temperatures $T_8 = T_9$ (one by the regenerator, the other by the compression), mix together and pass through the high temperature regenerator. The gas is heated in the heater from temperature T_6 to the maximum temperature T_3, expanding in the turbine up to point 4, then cooled first in the high temperature regenerator, then in the low temperature regenerator, down to temperature T_7: from where the cycle restarts.

Dividing the gas flow in point 7 has a similar effect to that of regeneration in the steam cycles. In the high temperature regenerator, the heat capacities of the two flows are not very different because of the elevated temperature, significantly higher than the critical temperature, which tends to lessen the effects of real gas (see Sect. 2.2).

Figure 4.13a compares the efficiency results of cycles with total condensation, with partial condensation and with completely supercritical Brayton cycles. The efficiency of cycles with partial condensation and split compression reach values close to 40 %. When $r_C = 4$, the efficiency of the cycle with partial condensation (shown on plane T-S in Fig. 4.13b) is about 0.39, compared with a value of 0.34 for the cycle with total condensation and compression entirely in the liquid phase.

Thermodynamic analysis has shown that to obtain the maximum thermodynamic benefits connected with the effects of real gas, it is necessary to make thermodynamic cycles with heat rejection and compression phase close to the critical point (see Fig. 4.8). In general, thermodynamic performance increases considerably if there is also present in the cycle a condensation phase (see Fig. 4.13), with expansion in the region of the supercritical gas.

Figure 4.14 represents the efficiency of the supercritical Brayton cycles as the minimum temperature T_1 varies. The temperature ratio T_3/T_1 is fixed at a value of 3 (ideal efficiency of 0.667). As we can see, even from a temperature T_1 of 40 °C (reduced temperature of 1.03), the positive effects of the real gas on the thermodynamics begin to diminish, in particular at low compression ratios.

For many aspects, carbon dioxide is effectively a good working fluid: its thermodynamic properties are well known, it is not toxic and it is abundant and

Fig. 4.14 Efficiency of supercritical Brayton cycles with carbon dioxide, as a function of the compression ratio and for different values of the temperature T_1 at the start of compression

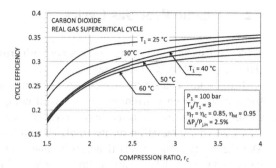

inexpensive. The compatibility of carbon dioxide with materials should not be underestimated: it was thoroughly investigated during the development projects for the English nuclear reactors of Magnox and AGR [13], which were cooled with carbon dioxide. Serious problems of corrosion of the steel in the steam generators by carbon dioxide at 400 °C occurred in the Magnox-type reactor at Latina (Lazio, Italy).[2] On the other hand, the AGR reactors, made with stainless steel components, reached, in the few examples built, temperatures of 650 °C for the heat-carrying fluid (carbon dioxide), with operating pressures of 30–40 bar.

Studies into carbon dioxide corrosion are currently in progress, since the problem of carbon dioxide compatibility with construction and stainless steels at high, even supercritical, pressures is relevant today for the oil industry in oil pipelines and is fundamental for the technologies of carbon capture and storage. The presence of contaminants, like water, sulphur dioxide SO_2 and nitrogen dioxide NO_2, can exacerbate the phenomena of corrosion.

From a strictly thermodynamic point of view, one disadvantage is its critical temperature of 31.06 °C, which, in order to exploit properly all the benefits of the effects of a real gas, requires a low temperature cooling fluid, for example, water at 10–20 °C. With carbon dioxide thermodynamic cycles, even the evaporation towers could be inadequate if the climate is hot and humid.

Organic compounds offer a great variety of fluids, with highly varied critical temperatures, and careful selection may find working fluids with critical temperatures better suited than carbon dioxide. For example, critical temperatures around 30 °C for cycles producing just electricity, in places where water for cooling is widely available, and critical temperatures of 50–80 °C for CHP generation. The possibility of resorting to appropriate mixtures of organic fluids would also permit a continuous variation in the critical temperature, truly optimising the working fluid with respect to the sink temperature [14]. These aspects will be discussed in Sect. 4.2.

---

[2]A reactor with a nominal electrical power of 200 MW. Operative between 1963 and 1987. In 1969, the significant oxidisation of the mild steel in the steam generator by carbon dioxide forced them to reduce the maximum temperatures of the carbon dioxide to 360 °C, reducing the useful power by about 20 %.

4.2 Organic Real Gas Cycles

Table 4.1 lists several organic compounds with critical temperatures between 25 °C (HFC 23) and 150 °C (n-butane).

These can be used in Brayton cycles with notable real gas effects (and, in some cases, when convenient, cycles with compression in the liquid phase and expansion in the supercritical gas phase) that can meet various operating needs: either just the generation of electricity or combined electric and heat generation. The thermodynamic cycles that can be made are completely analogous to those described in Sect. 4.1 for carbon dioxide, except for the differences in the molecular complexity (the parameter σ described in Sect. 2.5) and the molar mass of the working fluid: as the molecular complexity rises, there is less cooling of the gas in the turbine during the expansion, having set the expansion ratio, with a consequent increase in the heat generated per unit of useful power; high molar masses correspond to moderate differences of enthalpy in compression and expansion, with the possibility of making single-stage turbomachinery that is very conservative from the point of view of mechanical stress on the blades.

Due to the high density of the fluid (a characteristic which is obviously true also for carbon dioxide and common to all those cycles with values of T_1 and P_1 close to the critical point), the dimensions of the turbomachines are limited, considering the high power levels (for example, average diameters of 0.2 m at 20–30 thousand revs for power of around one MW), or, alternatively, positive displacement engines can be made for lower power levels (10–100 kW).

However, the thermal stability limits the maximum operating temperatures of the organic fluids (see Sect. 3.2). Among the compounds listed in Table 4.1, the fluorocarbons HFC-23, HFC-125 and HFC-134a are, according to specific tests, unaffected by the temperature, at least up to 350–400 °C [14]. The hydrocarbons (ethane, propane and butane) are very unlikely to be stable at temperatures higher than 300–350 °C. With reference to HFC-125, Fig. 4.15 compares the efficiency of condensation cycles and totally supercritical cycles with carbon dioxide. In the examples discussed below, the minimum temperature T_1 is assumed to be 40 °C, considering the possibility of the surrounding air as cooling fluid. The advantage of being able to use a fluid that permits compression in the liquid phase is evident, both in terms of efficiency and the maximum pressures of the cycle.

Figure 4.16 compares two cycles with condensation at $T_1 = 40$ °C, operating with HFC-125. The cycle in Fig. 4.16a is a cycle with total condensation; that in Fig. 4.16b is with partial condensation. The maximum temperature T_3 is equal to 400 °C.

The partial condensation and the split compression significantly reduce the irreversibilities of the heat exchange in the regenerator (reducing the temperature differences between the hot fluid and the cold fluid, see Fig. 4.16c,d), with notable increases in the cycle efficiency.

Table 4.1 Critical temperature and pressure as well as molar mass for several organic compounds

Fluid	Critical temperature (°C)	Critical pressure (bar)	Molecular weight
Carbon dioxide	30.97	73.74	44.01
HFC 23[a]	25.82	48.36	70.014
Ethane	32.17	48.72	30.07
Sulphur hexafluoride	45.57	37.6	146.056
HFC 125[b]	66.02	36.15	120.022
R 218[c]	71.95	26.8	188.02
HFC 32[d]	78.11	58.05	52.024
Propane	96.68	48.72	44.097
HFC 134a[e]	101.11	40.59	102.032
Octafluorocyclobutane (C_4F_8)	115.22	27.78	200.031
Butane	151.97	37.96	58.123

[a]Trifluoromethane
[b]1,1,1,2,2-Pentafluoroethane
[c]Octafluoropropane (C_3F_8)
[d]Difluoromethane
[e]1,1,1,2-Tetrafluoroethane

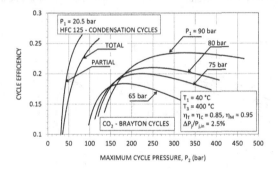

Fig. 4.15 Efficiency for cycles with HFC-125 and carbon dioxide as working fluids, as a function of the maximum pressure

Multi-Component Working Fluids

The possibility to mix working fluids with different critical temperatures will, in principle, enable the realisation of closed cycles with significant real gas effects for any minimum temperature T_1, not forgetting, though, the technological limit on the maximum temperature T_3 and bearing in mind all the considerations in Sect. 3.2.

Figure 4.17a represents several bubble and dew curves for mixtures of HFC-23 and HFC-125 and the envelope line of the points at maximum pressure as the fraction y_1 (molar fraction of HFC-23) varies: the critical temperature passes steadily from the value relative to pure HFC-23 (about 26 °C) to the value of

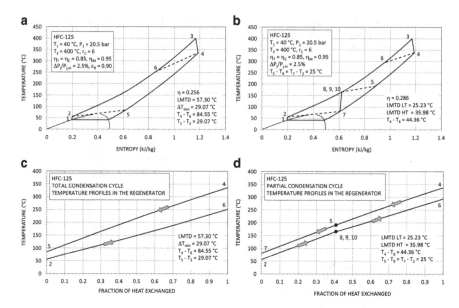

Fig. 4.16 Cycles with condensation and expansion in the supercritical gas phase, with HFC-125 as the working fluid on the temperature–entropy plane and diagrams of the heat exchange of the regenerator

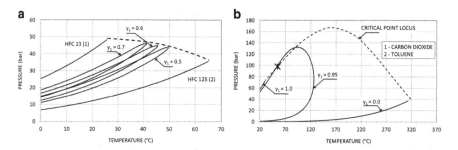

Fig. 4.17 Bubble and dew point curves for mixtures with different compositions. The equation used for the calculations is (2.5a) with the mixing rules (2.30). (**a**) Mixtures HFC-23 + HFC-125. The coefficient $\delta_{1,2} = -4.662409\,10^{-3}$ was obtained with a regression of the experimental data reported in [15]. (**b**) Mixtures carbon dioxide + toluene [16]

HFC-125 (66 °C). The figure traces the liquid–vapour equilibrium curves for three compositions: $y_1 = 0.7, 0.6$ and 0.5. Assuming $T_1 = 40\,°C$, the mixture with $y_1 = 0.7$ enables supercritical Brayton cycles to be made with real gas effects; the mixtures with $y_1 = 0.6$ and 0.5 already enable the design of cycles with condensation. Condensation in non-azeotropic mixtures, though, is not isothermal: for instance, in the case of the mixture with $y_1 = 0.5$, the temperature glide (at the pressure of 37.5 bar) is 4.6 °C (the temperature during the condensation varies from 44.6 °C, in correspondence with the dew point, to 40.0 °C in correspondence with the bubble point).

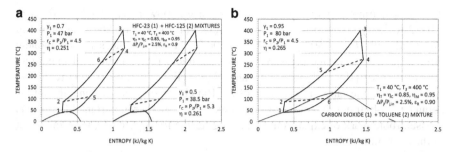

Fig. 4.18 Examples of supercritical Brayton cycles, with compression in the liquid phase for the mixtures. (**a**) Mixtures of HFC-23 + HFC-125. (**b**) A mixture of carbon dioxide + toluene [16]

Figure 4.17b reports the liquid–vapour equilibrium curves for mixtures of carbon dioxide and toluene (some physical properties of toluene are in Table 3.3). In this case, too, the critical point of the mixtures varies steadily (even though in an evidently non-linear fashion) between the values corresponding to the two pure fluids. Figure 4.17b shows the coexistence curve of the phases for the mixture with molar composition $y_1 = 0.95$ (carbon dioxide) and $y_2 = 0.05$ (toluene): the critical point of the mixture has a temperature around 55 °C and a pressure of about 98 bar; the point of maximum pressure (on the dew line, at about 132 bar) is at a temperature of 92 °C.

Figure 4.18a reproduces, on its thermodynamic plane T-S, two cycles with mixtures HFC-23 + HFC-125, while a gas cycle with compression in the liquid phase with carbon dioxide + toluene is shown in Fig. 4.18b.

Mixtures, therefore, constitute a valuable instrument for steadily varying the critical point of the working fluid, in such a way as to make Brayton cycles with real gas effects, for different applications with various minimum temperatures T_1. The variations in critical temperature and pressure with the composition of the mixture are not generally linear and experimental data and reliable calculation models are required to predict them. From a strictly thermodynamic point of view, there is a substantial similarity between totally supercritical Brayton cycles and condensation cycles, the choice depending on the applications (power levels, type of cooling fluid available etc.). The mixtures have heat exchange coefficients in condensation (and in evaporation) inferior to those of the corresponding pure fluids and this, in principle, favours (in the case of mixtures as working fluids) the use of totally supercritical Brayton cycles. The absence of two-phase flows, in certain particular cases, may prove beneficial in the design stage (for example, in space applications, in the absence of gravity). The choice of the mixture components has a direct influence on the design characteristics: for example, the cycle in Fig. 4.18b, has a minimum pressure of 80 bar and a maximum pressure of 360 bar and regeneration takes place partly with the two-phase flows and the glide temperature in condensation is high.

The Brayton cycles with real gas are characterised by large quantities of thermal power to be regenerated per unit of useful power, like in Brayton cycles with ideal gas (see Sect. 1.7). For example, for the cycle in Fig. 4.16a, $\dot{Q}_R / \left(\dot{W}_T - \dot{W}_C \right) = 5.6$

with a heat transfer parameter per unit of useful power $\dot{Q}_R/\Lambda\left(\dot{W}_T - \dot{W}_C\right) =$ 0.0972 K^{-1}. By comparison, a Rankine cycle with toluene (of the type shown in Fig. 3.9a) with $T_1 = 40\,°C$ and $T_4 = 400\,°C$, assuming the same component efficiencies and the same pressure losses, has, under the conditions of maximum efficiency, $\dot{Q}_R/\left(\dot{W}_T - \dot{W}_C\right) = 1.15$ and $\dot{Q}_R/\Lambda\left(\dot{W}_T - \dot{W}_C\right) = 0.0297$ K^{-1}. The isentropic volume ratio of expansion of the Rankine cycle with toluene, though, is about 670 (140 times that of the cycle in Fig. 4.16a); another important difference is the isentropic work of expansion: 292 kJ/kg for the Rankine cycle, four times that of the cycle in Fig. 4.16a.

The gas cycles with real gas can be a valid alternative to closed Brayton cycles with an ideal gas. For example, the cycle in Fig. 1.42 (Brayton cycle with helium, at $T_1 = 20\,°C$ and $T_3 = 850\,°C$, with intercooling) has an efficiency of 40 %; the cycle with carbon dioxide in Fig. 4.13b has the same efficiency at a temperature $T_3 = 600\,°C$ (with relatively modest component efficiency and significant pressure losses).

The great thermal energy recovered in the regenerator per unit of useful power requires a demanding heat exchanger, at not particularly high maximum temperatures (400–700 °C), but with high operating pressures (in the order of 100 bar, in the range 100–300 bar). This requires the use of compact heat exchangers, made with pipes and plates of small diameter and width, to keep the matrix of heat exchange compact and light, but resistant to pressure differences even at high temperatures. A very promising technology appears to be that of PCHEs (printed circuit heat exchangers), which consist of channels of 2 mm diameter, chemically etched in metal sheets which are then diffusion bonded to create solid blocks, which are then welded together to make the complete heat exchanger [9, 17, 18].

Controlling the power in Brayton cycles with real gas may not be possible by simply varying the pressure levels in the system (as in Brayton cycles with ideal gas) and the control system should be designed appropriately, especially in the case of plant layouts that envisage split compression.

Exercises

4.1. Liquefied natural gas (LNG) is a natural gas cooled to $-162\,°C$. At this temperature, the natural gas, at room pressure, condenses and becomes liquid. When it reaches a liquid state, natural gas occupies 600 times less space than in its gaseous state, making it possible to transport it over long distances and, under the form of LNG, it can easily be shipped at competitive prices from the producing countries to those in the world with the greatest demand. LNG is odourless, colourless, noncorrosive and non-toxic, with a density less than half that of water. It is expected that in years to come, the volumes of LNG being sold will be twice the current level.

Once it reaches its destination, the LNG is compressed, re-gasified and emitted into the network. An interesting option consists in using the cold LNG as a heat sink for power cycles and exploiting its physical exergy to produce electricity. One example, with closed cycle and perfect gas, is discussed in Exercise 1.11. Here we shall discuss several thermodynamic condensation cycles with carbon dioxide and real gas Brayton cycles.

We imagine disposing of heat at a high temperature (generated by burning fuel), in such a way as to increase the conversion efficiency, increase the useful work per unit of re-gasified LNG and reduce the costs per unit of installed power. We suppose that the LNG is available as a supercritical gas at 70 bar (a typical value for long-distance gas pipelines).

The maximum useful power $\dot{W}_{\text{optimal}}$ obtainable from heating the LNG from the minimum temperature of $-160\,°C$ up to the ambient temperature of $25\,°C$ can be derived from (B.4), assuming $\dot{S}_G = 0$ and with $\dot{E}_Q = 0$ and $T_0 = 298.15$ K this gives $\dot{W}_{\text{optimal}} = -443.196$ kW/kg/s. Having thermal power available at the temperature T_3 (higher than the temperature T_0 of the environment), (B.4) gives as a result:

$$\dot{W}_{\text{optimal}} = \frac{T_3}{T_0}\left[(\dot{E}_{\text{out}} - \dot{E}_{\text{in}}) - \left(1 - \frac{T_0}{T_3}\right)\dot{m}_{\text{LNG}}(H_{\text{out}} - H_{\text{in}})\right]$$

which, for $T_3 = 600\,°C$, gives $\dot{W}_{\text{optimal}} = -2825.56$ kW/kg/s.

Evaporation of the liquified gas may take place at the condenser of a supercritical cycle with carbon dioxide, but there needs to be a compromise between the high cycle efficiency (which would require the lowest possible condensation temperature) and the maximum use of the physical exergy of the LNG (which would require an elevated final temperature, only possible from a high condensation temperature, that is, a modest cycle efficiency). The Specific Power Performance Parameter (SPP, in MW/kg/s of LNG, see [19]), therefore, tends to grow with the condensation temperature, while the cycle efficiency increases regularly with the drop in the condensation temperature. Figure 4.19a has the results of efficiency and SPP for cycles with carbon dioxide. These are supercritical Rankine cycles with carbon dioxide (of the same type as those discussed previously, see Fig. 4.11c, but with condensation temperatures T_1 that are significantly lower than the critical temperature). The results relate to the hypothesis of using sea water in a sea water heater to preheat the carbon dioxide from temperature T_2, close to the condensation temperature (for example, $-50\,°C$), up to a temperature (for example, $20\,°C$) close to room temperature. The efficiencies of the components and the parameters assumed for the calculations are those in Table 4.2.

At temperature $T_1 = -50\,°C$, for example, corresponding to a maximum pressure $P_2 = 150$ bar, the cycle efficiency is around 0.52 and the SPP is 0.36. Figure 4.19b represents a thermodynamic cycle in the temperature–entropy plane. The simple cycle, adapted to the partial use of heat extracted from sea water, has good efficiency and the plant configurations are not too complex.

There is more detailed discussion of the performance of carbon dioxide condensation cycles, in the case of the particular application considered here, in [19], where we also talk about cycles with multiple condensation levels and compound cycles (of the type in Fig. 4.12), at high and low pressures.

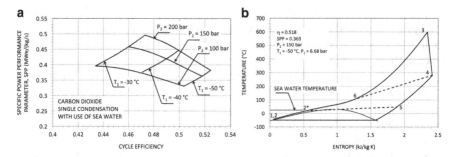

Fig. 4.19 Simple condensation cycles with carbon dioxide and with partial use of sea water. (**a**) Cycle efficiency and specific power output SPP. (**b**) Example of the cycle on the thermodynamic plane T-S

Table 4.2 Efficiencies of components and parameters assumed for the carbon dioxide cycles of Fig. 4.19

Turbine inlet temperature, T_3	600 °C
Condensation temperature, T_1	from −50 °C to −30 °C
Turbine adiabatic efficiency, η_T	0.90
Pump efficiency, η_C	0.70
Regenerator efficiency, ϵ_R	0.90
Logarithmic temperature difference in the condenser, Λ_C	50 °C
Mechanical efficiency, η_M	0.985
Alternator efficiency, η_A	0.98
Fractional pressure drops on each heat exchanger, $\Delta P_j / P_{j,in}$	2.5 %

In any case, the supercritical LNG, re-gasifying at variable temperatures, would be difficult to match with the condensation of a Rankine cycle, unlike the Brayton cycles, in which the heat is discharged as the temperature drops.

4.2. The ideal gas closed cycle has the potential to use almost completely the cooling capacity of supercritical LNG available at −160 °C and at supercritical pressures (for example, 70 bar; see Exercise 4.1). An alternative to using the closed Brayton cycle with perfect gas is the closed real gas Brayton cycle which, as we have seen in this section, is characterised by a drastically reduced work of compression that, to some degree, makes these cycles similar to the Rankine cycles. The significant reduction in the compression work guarantees good cycle efficiency even at less than high maximum temperatures. Important effects of the real gas on the compression become manifest, though, only in proximity of the critical point (for example, when $T_{1,r} = 0.98 - 1.01$), and this is the reason why, in usual applications, the permanent gases (air, nitrogen, helium) do not show any appreciable effects of the real gas at ordinary temperatures (see Fig. 4.2b). The situation is different if the closed gas cycle operates with minimum cryogenic temperatures, as in the case of the re-gasification process of LNG. Table 4.3 lists some of the substances that, in principle, could be used to make real gas Brayton cycles at temperatures T_1 close to that of the LNG.

Table 4.3 Critical temperature and pressure as well as molar mass for several working fluids in cryogenic Brayton cycles

Fluid	Critical temperature (K)	Critical pressure (bar)	Molecular weight
Nitrogen	126.2	33.98	28.014
Air	132.52	37.66	28.96
Argon	150.86	48.98	39.948
Oxygen	154.58	50.43	31.999
Methane	190.56	45.99	16.043

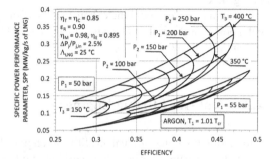

Fig. 4.20 Specific Power Performance Parameter (SPP, useful power per unit mass flow of vaporised LNG) for argon cycles and cycle efficiency for different minimum pressures and maximum pressures and temperatures

The modest compression work due to the relative incompressibility of the real gas is accompanied, though, by a fall in the end of compression temperature T_2 and, consequently, by a modest temperature T_5 (see Fig. 4.11a,b). This has the negative effect (in the case of the particular application considered here) of reducing the SPP specific power parameter.

Figure 4.20 reports the SPP Parameter (SPP) and the efficiency of real gas Brayton cycles with argon ($T_1 = 1.01 T_{cr}$) under different operating conditions. Passing from a minimum pressure P_1 of 50 bar to a P_1 of 55 bar, the cycle efficiency increases by an average of 2–3 points, but the SPP parameter drops significantly. For example, in correspondence with $T_3 = 350\,°C$ and $P_2 = 200$ bar, when $P_1 = 50$ bar the efficiency is 0.446 and the SPP is 0.289 MW/kg/s; when $P_1 = 55$ bar, we have a cycle efficiency of 0.46 and an SPP of 0.177 MW/kg/s (60 % lower): it is usually necessary to make a compromise between high efficiency values and acceptable values of the SPP parameter.

It is worth noting how, at even relatively modest T_3 temperatures (150–200 °C), the cycle efficiency is interesting. For example, it is the same at 0.316 with $P_1 = 50$ bar and $T_3 = 150\,°C$, but with SPP = 0.123 MW/kg/s. This would permit the use of easily recovered low-grade heat as heat source, allowing a real gas Brayton cycle to be used (with small expansion ratios, for example, equal to 3), in place of an ORC. There is more detailed discussion of the potential of real gas Brayton cycles associated with the re-gasification of LNG in [20].

In this case, too, we should not overlook the possibility of mixing different gases (for example, argon and nitrogen) in such a way as to obtain a variation in the critical point and, thereby, broaden the choice of operating conditions.

4.3. Still with reference to the integration of real gas Brayton cycles into the re-gasification process of LNG, we consider the case of a "binary" cycle obtained with a commercial gas turbine of 4,600 kW (Mercury 50, a recuperated gas-turbine generator set) and with an argon "bottoming" cycle.

The characteristics of the gas turbine are the following: exhaust mass flow 63,700 kg/h, heat rate 9,351 kJ/kWh and exhaust temperature 377 °C. We consider a recovery cycle with argon with $T_1 = 1.01T_{cr}$, $T_3 = 350$ °C, $P_1 = 50$ bar and a compression ratio r_C of 4.5. The characteristics of the components are those reported in Fig. 4.20.

Assuming we can cool the exhaust gases of the gas turbine down to 100 °C, we get a power of the real gas Brayton cycle of 2,360 kW, an overall efficiency of 0.58 and an SPP of 0.90.

4.3   Real Gas Stirling Engines

The thermodynamics of ideal gas Stirling engines have been discussed in Sect. 1.8. In modern Stirling engines, the working fluid is typically helium or hydrogen, because of their good heat exchange properties, see Sect. 1.7, but there are severe sealing problems. Air has been proposed repeatedly for engines with a simple design, low rev and with modest specific power [21]. In any case, the working fluid employed operates in thermodynamic conditions that oblige it to behave as an ideal gas.

In the case of Stirling engines, as in Joule–Brayton engines, the use of real gas brings about a noticeable increase in the specific work and guarantees a discrete efficiency even at maximum operating temperatures that are inferior to those typical of perfect gas engines.

The Malone[3] engine was the first engine, regenerative and with external combustion, similar to the hot-air Stirling engine, which employed water as its working fluid in place of air. In the Malone engine, the very low compressibility of the fluid led to very high maximum pressures (800–1,000 bar), with very low rotational speed

---

[3]John Fox Jennens Malone, an Englishman, was born at Wallsend on Tyne in 1880 and died in 1959 at Newcastle upon Tyne. At eighteen years of age he entered the merchant navy and served for fourteen years. In the 1920s, he began to test heat engines with liquids as the working fluids and, in 1927, he completed a small water engine of 50 HP. Malone wrote in 1931 that "Trials by three different independent engineers gave 27 % indicated efficiency" [22] (a value that was two to three times greater than that of the steam engines used at the time on ships and locomotives). In 1932, Malone founded the Malone Instrument Co. Ltd, but his engine with liquid as its working fluid enjoyed no success, thanks also to the rapid growth of the high powered steam turbines and the internal combustion engines. In a letter of 1939, Malone gave vent to his bitterness: "there is one fact, a study of liquids as mediums in thermodynamics will teach an engineer more about the art of thermodynamics than all the universities on earth, or the memory men who infest them, and knowledge for knowledge's sake is better than their parasitical life" [23].

(30–300 rpm) and with small swept volumes of the pistons. Malone tried out various working fluids, among which is carbon dioxide [24].

The approach taken by Malone (followed later by other authors [25, 26]), in principle, makes it possible to create heat engines with liquid as a working fluid, positioning the expansion in proximity of the critical point, in Region 6 of Fig. 2.1, and the compression in Region 5 of Fig. 2.1 as, for example, the thermodynamic liquid cycle E in Fig. 4.1.

In fact, the useful power of a Stirling engine may be calculated as $\dot{W} = -P\,(dV_C/dt + dV_E/dt) = -P\,dV/dt$, with V representing the total instantaneous volume of the engine. The pressure $P = P(t)$ over time is not easy to calculate (see Sect. 1.8.1). Qualitatively, assimilating the engine to a closed system, at an appropriate average temperature, the pressure can be easily derived from the single equation (A.1) of the balance of mass: $dM_{tot}/dt = 0$, with $M = \rho V(t)$. Or:

$$\frac{d\rho}{dt}V + \rho\frac{dV}{dt} = 0 \quad \text{with}$$

$$\frac{d\rho}{dt} = \left(\frac{\partial\rho}{\partial P}\right)_T \frac{dP}{dt} + \left(\frac{\partial\rho}{\partial T}\right)_P \frac{dT}{dt} \quad \text{in which}$$

$$\frac{dT}{dt} = \left(\frac{\partial T}{\partial P}\right)_S \frac{dP}{dt}$$

$$= (-1)\frac{T}{\rho^2 C_P}\left(\frac{\partial\rho}{\partial P}\right)_T \frac{dP}{dt} \quad \text{for an adiabatic expansion or compression or}$$

$$\frac{dT}{dt} = 0 \quad \text{for a isothermal expansion or compression}$$

So, for instance, with reference to an isothermal transformation, we get

$$\left(\frac{\partial\rho}{\partial P}\right)_T \frac{dP}{dt}V + \rho\frac{dV}{dt} = 0 \quad \text{or}$$

$$\frac{dP}{dt} = (-1)\rho\frac{dV}{dt}\frac{1}{V\left(\frac{\partial\rho}{\partial P}\right)_T}$$

The tendency of the engine pressure is, therefore, strictly correlated to the compressibility $(\partial\rho_P) = (\partial\rho/\partial P)_T$. For a liquid under normal conditions (with $T = T_K$ close to room temperature) $(\partial\rho_P)$ is very small (null, if the liquid was rigorously incompressible); on the contrary, for a liquid with $T = T_H$ close to the critical temperature, $(\partial\rho_P)$ is significantly greater than zero. The result is that even a working fluid in the liquid phase can guarantee finite variations of volume (modest, but not infinitesimal) with similarly finite pressures (although tending to be very high), with a net useful work on the cycle.

Fig. 4.21 Reduced density ρ_r
as a function of the reduced
pressure P_r in
correspondence with different
reduced temperatures T_r. The
data refer to water [27]

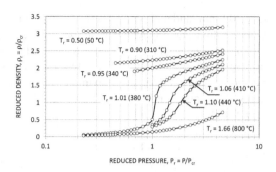

Figure 4.21 reports reduced density values ρ_r, at different reduced pressures and
reduced temperatures. Although the data in the figure refers to water, the trend of
the reduced density as T_r and P_r vary, is, by the Law of Corresponding States (see
Sect. 3.2), of general validity. Figure 4.21 clearly shows the great variations in the
density as pressure varies in proximity to the critical temperature.

In the case of an engine with liquid as its working fluid, the variations of volume
V, necessarily modest, imply small values of the expansion swept volume $V_{SW,E}$,
small values of the ratio $\kappa = V_{SW,C}/V_{SW,E}$ and relatively large ratios $\zeta = V_D/V_{SW,E}$.
Of greater interest for the application point of view, is an engine that uses a gas, but
with significant real gas effects. In such a case, reducing the average compressibility
of the fluid enclosed within the volume $V = V(t)$ of the engine, compared to the
case with ideal gas, for similar volume variations, increases the average pressure
anyway, with a significant increase in the work parameter $W^* = W/P_{max}V_T$.

The thermodynamic model described in Sect. 1.8.1 will now be applied to
Stirling cycles with real gas and ideal gas. The design parameters are the usual ones:
the phase angle ϕ, the ratio between the maximum temperature and the minimum
temperature $\tau = T_H/T_K$, the ratio between the swept compression volume and the
swept expansion volume $\kappa = V_{SW,C}/V_{SW,E}$, the ratio between the dead volume and
the swept expansion volume $\zeta = V_D/V_{SW,E}$, the number of revs N, the efficiency
of compression η_C and of expansion η_E (defined according to (1.62) and (1.63)),
the fractional temperature difference in the regenerator ϵ defined by the condition
(1.60) and by (1.61). For the calculations carried out below, the values reported in
Fig. 4.22a have been assumed for the design parameters.

The results (valid within the limits of the calculation model used) for the
comparison of engine performance are the work parameter $W^* = W/P_{max}V_T$ and
the (indicated) efficiency of the engine $\eta = W/\oint \dot{Q}_H dt = W/Q_H$ (see Sect. 1.8.1).

The geometric parameters of the engine listed earlier also define the ratio
between the dead volume V_D and the total volume V_T:

$$\frac{V_D}{V_{SW,E}} = \zeta \quad \text{and}$$

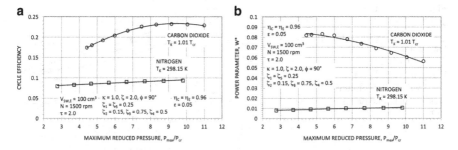

Fig. 4.22 Comparison between the thermodynamic performance of Stirling engines with nitrogen and with carbon dioxide. (**a**) Cycle efficiency as a function of the maximum reduced pressure of the cycle. (**b**) Work parameter W^*

$$\frac{V_T}{V_{SW,E}} = 1 + \kappa + \zeta \quad \text{so,}$$

$$\frac{V_D}{V_T} = \frac{\zeta}{1 + \kappa + \zeta}$$

Then:

$$\frac{V_D}{V_{SW,E}} = \zeta$$

$$= \frac{V_{CL,C}}{V_{SW,E}} + \frac{V_K}{V_{SW,E}} + \frac{V_R}{V_{SW,E}} + \frac{V_H}{V_{SW,E}} + \frac{V_{CL,E}}{V_{SW,E}}$$

$$= \sum_{i=1}^{5} \zeta_i$$

and, having set the four values of ζ_i, from the previous equation, we can derive the fifth. In the calculations that follow below, the fractions ζ_i that were assumed are listed in Fig. 4.22a.

The cycle efficiency with carbon dioxide (see Fig. 4.22a) reaches a maximum value of around 23 % in correspondence with a maximum pressure of about 9 times the critical pressure. The maximum efficiency, in the range of maximum reduced pressures considered, is 2.3 times greater than the value of the efficiency of cycles with nitrogen (ideal gas). The maximum values of the work parameter W^* for carbon dioxide (see Fig. 4.22b) are reached with maximum pressures between 4.5 and 6 times the critical pressure and are higher than the corresponding values for nitrogen by about 8.3 times.

Compared to the nitrogen cycles, the carbon dioxide cycles have optimal efficiency, although the temperature T_H, having assumed a ratio $\tau = T_H/T_K = 2$, is

Fig. 4.23 Pressure–volume
diagrams for carbon dioxide
cycles and nitrogen cycle. For
the carbon dioxide cycle,
$T_K = 1.01T_{cr}$ and for the
nitrogen cycle, $T_K =$
298.15 K. The maximum
pressure in both cases is
442 bar

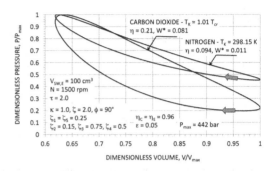

Fig. 4.24 Efficiency losses
and cycle efficiency as
maximum reduced pressure
varies. The working fluid is
carbon dioxide. The
temperatures T_K and T_H and
the parameters used for the
calculations are the same as
those used for Fig. 4.22a

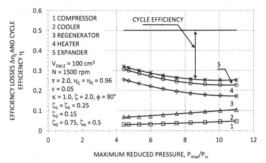

relatively modest (341 °C for cycles with carbon dioxide and 323 °C for cycles with nitrogen). This is a consequence of the effects of real gas which, having localised the temperature T_K and the minimum pressure in proximity to the critical point of carbon dioxide, are responsible for a large pressure variation $(P_{max} - P_{min})/P_{mean}$: 1.51 in the case of the carbon dioxide cycle of Fig. 4.23 and 0.81 in the case of the nitrogen cycle.

The existence of the maximum efficiency value of Fig. 4.22a for the carbon dioxide cycles is explained by entropic analysis, which enables us to calculate the efficiency losses $\Delta\eta_j = T_0\dot{S}_{G,j}/\dot{Q}_{in}$ reported in Fig. 4.24 (for the calculation, see Exercise 1.3, Appendix A.4 and C.3). The maximum efficiency value is a consequence, firstly, of the substantial and rapid diminishing of thermodynamic losses in the regenerator as the maximum pressure increases and then, by ever-increasing maximum pressures, of the prevalence of compressor and cooler losses.

So, the considerations are similar for the real gas Stirling engines to those of the real gas Brayton cycles: (1) they have good efficiency at high specific useful power; (2) there is the possibility of using multicomponent working fluids in order to control the critical temperature and meet, in principle, various operating needs; (3) good efficiency is obtained even with not particularly high maximum temperatures (with significant technological advantages in the choice of materials); and on the downside, (4) the maximum pressures which interest us are significantly higher than the critical pressure and, therefore, tend to be rather high (several hundred bars). On the other hand, the high pressures, the temperature close to the critical temperature and the compressibility factors Z well below the unit, can prove beneficial and the ratio $\dot{W}_f / \dot{Q}$, between the power spent in overcoming the pressure losses and the thermal power exchanged in a heat exchanger (see (1.40) and the relative discussion), can also turn out to be ten times lower than the typical value for an ideal gas.

References

1. Krasin AK, Nesterenko VB (1971) Dissociating gases: a new class of coolants and working substances for large power plants. Atom Energ Rev 9(1):177–194
2. Angelino G (1979) Performance of N_2O_4 gas cycles for solar power applications. Proc Inst Mech Eng 193(1):313–320
3. Anonymous (1956) Calder Hall power Station. The Engineer, 5 October, pp 464–468
4. Feher EG (1967) The supercritical thermodynamic power cycle. In: Advances in energy conversion engineering. Intersociety energy conversion engineering conference, Miami Beach, FL, pp 37–44, 13–17 August
5. Angelino G (1967) Perspectives for the liquid phase compression gas turbine. J Eng Power Trans ASME 89(2):229–237
6. Angelino G (1967) Liquid-phase compression gas turbine for space power applications. J Spacecraft Rockets 4(2):188–194
7. Angelino G (1968) Carbon dioxide condensation cycles for power production. J Eng Power Trans ASME 90(3):287–295
8. Angelino G (1971) Real gas effects in carbon dioxide cycles. Atomkernenergie (ATKE) 17(1):27–33
9. Dostal V, Driscoll MJ, Hejzlar P, Wang Y (2004) Supercritical CO_2 cycles for fast gas-cooled reactors. In: Proceedings of ASMETurbo Expo 2004. Power for land, sea, and air, Vienna, Austria, 14–17 June. Paper GT2004–54242
10. Hejzlar P, Pope MJ, Williams WC, Driscoll MJ (2005) Gas cooled fast reactor for generation IV service. Progr Nucl Energ 47(1–4):271–282
11. Dostal V, Hejzlar P, Driscoll MJ (2006) High-performance supercritical carbon dioxide cycle for next-generation nuclear reactors. Nucl Tech 154:265–282
12. Dostal V, Hejzlar P, Driscoll MJ (2006) The supercritical carbon dioxide power cycle: comparison to other advanced power cycles. Nucl Tech 154:283–301
13. Lee JC, Campbell J Jr, Wright DE (1981) Closed-cycle gas turbine working fluids. J Eng Power Trans ASME 103:220–228
14. Angelino G, Invernizzi C (2001) Real gas Brayton cycles for organic working fluids. Proc IME J Power Energ 215(1):27–38
15. Lim JS, Park JY, Lee BG (2000) Vapor-Liquid Equilibria of CFC alternative refrigerant mixtures: trifluoromethane (HFC-23) + difluoromethane (HFC-32), trifluoromethane (HFC-23) + pentafluoroethane (HFC-125), and pentafluoroethane (HFC-125) + 1,1-difluoroethane (HFC-152a). Int J Thermophys 21(6):1339–1349
16. Invernizzi C M, van der Stelt T (2012) Supercritical and real gas Brayton cycles operating with mixtures of carbon dioxide and hydrocarbons. Proc IME J Power Energ 226(5):682–693
17. Nikitin K, Kato Y, Ngo L (2006) Printed circuit heat exchanger thermal-hydraulic performance in supercritical CO_2 experimental loop. Int J Refrig 29:807–814
18. Min JK, Jeong JH, Ha MY, Kim KS (2009) High temperature heat exchanger studies for applications to gas turbines. Heat Mass Trans 46:175–186
19. Angelino G, Invernizzi CM (2009) Carbon dioxide power cycles using liquid natural gas as heat sink. Appl Therm Eng 29:2935–2941

20. Angelino G, Invernizzi CM (2011) The role of real gas Brayton cycles for the use of liquid natural gas physical exergy. Appl Therm Eng 31:827–833
21. Organ A J (2007) The air engine. Stirling cycle power for a sustainable future. Woodhead Publishing Limited, Cambridge
22. Anonymous (1993) John Malone and the invention of liquid-based engines. Los Alamos Sci 21:117
23. Sier R (2007) John Fox Jennens Malone. The liquid Stirling engine. L A Mair, Chelmsford
24. Malone JFJ (1931) A new prime mover. J Roy Soc Arts 79(4099):679–709
25. Allen PC, Knight WR, Paulson DN, Wheatley JC (1980) Principles of liquids working in heat engines. Proc Natl Acad Sci Unit States Am 77(1):39–43
26. Swift GW (1989) A Stirling engine with a liquid working substance. J Appl Phys 65(11): 4157–4172
27. Parry WT, Bellows JC, Gallagher JS, Harvey AH (2000) ASME international steam tables for industrial use. CRTD-Vol. 58. ASME Press, New York, NY

Chapter 5
The Binary Cycles

Today, the term "binary cycle" commonly refers to the organic fluid Rankine engines in geothermal systems, where a mass flow of geothermal brine is cooled in a recovery heat exchanger and the heat recovered is converted into electricity by means of an organic fluid engine (see Chap. 3, Sect. 3.6, and DiPippo [1]).

In general, though, the term "binary cycle" can cover any thermodynamic conversion system of heat into electricity with two different fluids involved in the process. In this sense, even the traditional combined gas turbine-steam cycle can be considered a binary system.

Out of the whole region of a general fluid's existence in the fluid phases, only the region enclosed by the limit curve or closely adjacent to it (see Fig. 4.1), permits the creation of thermodynamic cycles with high-quality thermodynamics[1] like the Rankine cycle with moderate superheating, illustrated by way of example in Fig. 5.1, cycle (a). As discussed in Sect. 1.6, the wide extend of the isothermal transformations, or almost isothermal, and the drastic contraction of the specific volumes following condensation, with a parallel containment of the work of compression, are at the basis of this excellent thermodynamic behaviour. Furthermore, the Rankine steam cycle enjoys an extreme adaptability to fuels and heat sources, but is unable to use heat efficiently at temperatures above 600–650 °C, the excess heat potential being lost as driving force of the heat transmission in the primary exchanger.

Just theoretically, a liquid substance at room temperature and with a critical temperature clearly exceeding the maximum temperatures technically allowed, would permit the creation of near-perfect thermodynamic cycles (see cycle (b) in Fig. 5.1). In reality, though, the exponential relation between the temperature and saturation pressure (see Fig. 2.3a) would quickly make the expansion ratio impractical as the temperature range in the cycle extends excessively, meaning that single-fluid solutions must make way for double or triple fluid solutions (binary or ternary cycles; see cycle (c) in Fig. 5.1).

---

[1] A good indicator of the thermodynamic quality of a cycle is the second-law efficiency, discussed in Exercise 1.2; see (1.22).

C.M. Invernizzi, *Closed Power Cycles*, Lecture Notes in Energy 11,
DOI 10.1007/978-1-4471-5140-1_5, © Springer-Verlag London 2013

Fig. 5.1 (**a**) Steam cycle.
(**b**) A hypothetical Rankine
cycle with liquid metal.
(**c**) Binary cycle liquid metal
steam

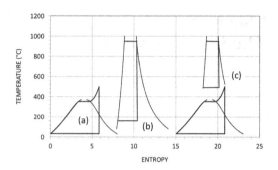

To this end, choosing a liquid other than steam would have the purpose of shifting the region of optimum use from the medium to low temperatures permitted by water vapour (below 300–320 °C) to high or very high temperatures (700–900 °C) permitted, for example, by metal vapours.

Thanks to the efforts of William Le Roy Emmet (1859–1941) and the General Electric Company, between 1923 and 1950, resolving tricky technical and technological problems, first there were built several prototypes of turbine and mercury boilers and, subsequently, several mercury and steam binary units.[2] In the mercury-steam binary cycles, the mercury is heated and vaporised in a mercury boiler and then expanded in a turbine. In a mercury-steam condenser, the exhausted mercury vapours—still at a high temperature during the condensation—heat and cause to evaporate the feedwater from the steam cycle. After the condensation, the liquid mercury returns to the preheater and then the boiler; the steam is then superheated by means of the gases produced by the combustion and originating from the mercury boiler and then sent to the steam turbine.

In 1923, an experimental mercury plant (of 1.8 MW) was built in the Dutch Point station in Connecticut, followed in 1928 by the first commercial cycle with a mercury unit of 10 MW and the steam turbine generator of 12.4 MW, for a total nominal power of 22.4 MW in the South Meadow station (Hartford, Connecticut). Between 1931 and 1933, General Electric built a plant at Schenectady (of 47 MWe) and another at South Kearney, New Jersey, of 45 MWe, both using pulverised coal. In 1949, the Schiller unit began operation, at Portsmouth, New Hampshire, with a total 43.4 MW (two mercury units and one steam unit). In 1949, operation also began at the General Electric power station in Pittsfield, Massachusetts, (of 21.6 MW) and a new unit at South Meadow (of 34.1 MW), [2, 3]. The Schiller unit was shut down in 1968, after 20 years of service.

According to date reported in [3, p. 578], the specific heat consumption of the South Kearny plant (which operated for 110,000 hours) was 14 % lower than that of

---

[2]The mercury vapours are extremely toxic, and the mercury does not wet the metal surfaces, creating major problems of heat exchange. The first mercury boilers were subject to tube plugging and failures, with massive corrosion of the tubes, made in low carbon steel, [2]. Water and air infiltration into the mercury circuits created further problems.

the steam power stations with a similar power in those years. The mercury turbine throttle pressures and temperatures in the last plants that came into operation in 1949 were (average values) 8.7 bar and 500 °C, respectively, with the exhaust conditions of about 0.15 bar and 260 °C. The reliability of the mercury vapour turbines (in single and double flow) was excellent. The rotation speed was between 720 and 1,200 rpm.

From 1950, mercury binary cycles were no longer built: the improvement in the operating conditions of steam cycles, their high power and significant economies of scale, as well as the high cost of mercury and the serious corrosion problems it created at high temperatures—never truly resolved—made the binary plants no longer financially viable. However, the study of thermodynamic cycles with liquid metals continued throughout the 1960s, in the United States, with the SNAP (System for Nuclear Auxiliary Power) programme and with the development of mercury units of 30–90 kW, with nuclear heat sources, to be used on space missions with expansion start temperatures of 650–700 °C. In those years, potassium was also considered and studied, and components and turbines were made with operating temperatures of 800 °C. No power plant with potassium has been built, though. In [4, 6] the two detailed projects for potassium-steam binary cycles in the 500–600 MW power range are reported. In [4] the proposal is for a molten salt nuclear fission reactor at high temperature (temperature of the reactor 982 °C, maximum temperature of the potassium cycle equal to 838 °C), with an efficiency of the combined system equal to 54.6 %. In [6] the potassium boiler is also the combustion chamber of a gas turbine.

Other organic and inorganic fluids were proposed as working fluids in topping cycles (even sulphur [5]). In [7], as an alternative to the use of mercury, binary cycles are proposed with diphenyl oxide[3] and steam. The cycle with diphenyl oxide, superimposed on the cycle with steam, has a maximum temperature of 400 °C. In the opinion of the paper's author the thermal decomposition of diphenyl oxide at 400 °C in normal carbon steel can be ignored. In [8] there is a description of a boiler steam plant with a superheating of the steam obtained from a eutectic mixture of diphenyl and diphenyl oxide at 382 °C and 7 bar. The plant operated for a year.

The metal liquids are suitable for creating high-temperature cycles positioned above a steam cycle. It is possible, though, to make binary cycles with the steam constituting the high-temperature section, as in the first binary cycles proposed, with the aim of preventing steam condensation at excessively low pressures (see [9]). A fairly recent, concrete example of this is the pilot plant described in [10]. An ordinary steam cycle uses the primary heat up to a condensation temperature of around 80 °C (0.5 bar) and releases the waste heat to an ammonia boiler at a pressure of about 35 bar. The ammonia cycle not only replaces the low-pressure section of the steam expander (the biggest, most mechanically stressed and costly) but also allows the power plant's waste heat to transfer, without further intermediary

---

[3]A heterocyclic organic compound, also called dibenzofuran, $C_{12}H_8O$. With $T_{cr} = 550.85$ °C, $P_{cr} = 36.34$ bar. With boiling temperature of 285.16 °C and a melting point at 81–85 °C.

circuits, to the dry towers with which the plant is equipped. The thermodynamic loss linked to the heat exchange in the condenser-evaporator is largely compensated by the lower condensation temperature that the system (unlike the ordinary steam system) ensures during the cold seasons. A second example of the use of organic fluid engines in steam power stations is described in [11]. The existing steam plants on the coast could use sea water, collected at a 100 m depth, which (along the Italian coasts) can be found at a perennial temperature of about 13 °C. These current plants are not able to use efficiently the potential of increased power given by using cooling water at low temperatures, because the passage section into the low-pressure elements of the turbine is insufficient. One plant scheme for resolving the problem could be to resort to organic fluid Rankine engines, which use the steam extracted during the expansion in the turbine and the condensate coming from the regenerators. The problems linked to the choice of bottoming cycles with fluids other than steam, on account of the low temperature, are the same as those that characterise the organic fluid Rankine engines, to which Chap. 3 was dedicated.

In Sect. 5.1 we shall discuss the binary cycles with liquid metals, which would permit a simplification of the technological problems connected with the use of materials at excessively high pressures and temperatures, typical of the supercritical steam cycles while, at the same time, significantly improving the conversion efficiency.

5.1 The Binary Liquid Metal–Steam Cycle

As we have said more than once, from a purely thermodynamic point of view, the best performances of a cycle are obtained when the heat is mainly introduced at the maximum operating temperature and the residual heat is rejected back into the environment at the lowest temperature: the saturated Rankine cycles are the practical example of this.

If a cycle is made on the limit curve, like type (b) in Fig. 5.1, the pressures vary exponentially with the temperature, and if the $\tau = T_H / T_C$ ratios are high, the expansion ratios are generally too high and difficult to handle. For example, in a mercury Rankine cycle with $T_C = 300$ K and $T_H = 900$ K ($\tau = 3$), the volumetric expansion ratio is in the order of 3×10^6, a value that cannot be handled in any real plant. So, the binary cycle with liquid metal and steam in which the waste heat of the unit at high temperature is released to the steam Rankine cycle below could be a solution.

Table 5.1 collects various thermo-physical properties for certain liquid metals of possible interest for use, maybe future, in energy conversion systems.

The low molecular complexity of liquid metals ($\sigma < 0$, Sect. 2.5) and the high critical temperature mean that once the evaporation temperature T_H is fixed, the reduced temperature (T_H / T_{cr}) is low and the contribution of preheating the liquid is small compared to the evaporation heat (see Sect. 2.4 and Fig. 2.3b), resulting in good quality thermodynamics for the Rankine cycle created (see also, for example, cycles (a) and (b) of Figs. 5.1, 1.14b, and 1.15). Regeneration, with the aim of

Table 5.1 Some physical properties of liquid metals for binary (or ternary) cycle applications

Fluid	M[a]	T_{cr} [b]	P_{cr} [c]	MT[d]	T_{vp} [e]	BT[f]
Lithium, Li	6.94	3,527	970	180.5	925.73	1,342
Sodium, Na	22.99	2,300	341	97.8	568.07	883
Magnesium, Mg	24.3	2,262	n.a.	650	764.36	1,107
Potassium, K	39.1	1,900	167	63.5	458.84	759
Rubidium, Ru	85.48	1,833	134	39.3	398.97	688
Cesium, Cs	132.9	1,775	117	28.5	384.72	671
Mercury, Hg	200.6	1,490	1,530	−38.4	195.61	356.7
Lead, Pb	207.2	5,127	n.a.	327.46	1,199.76	1,749
Bismuth, Bi	208.98	4,347	n.a.	271.4	1,094.16	1,564

[a]Molecular weight (g/mol)
[b]Critical temperature (°C)
[c]Critical pressure (bar)
[d]Melting temperature (°C)
[e]Saturation temperature at 0.02 bar (°C)
[f]Boiling temperature (°C)

reducing the entropic losses in the phase of preheating the liquid, is, therefore, not necessary for cycles with liquid metals.

The alkali metals constitute the first group in the periodic table of elements and they are lithium, sodium, potassium, rubidium and caesium. The first group also contains francium (Fr), which is very rare and radioactive (produced from the decay of actinium 227). All the alkali metals are highly reactive and are not to be found in their elementary state in nature, but they are fairly common in the form of compounds (Na $\approx$ 2.8 % in the lithosphere, K $\approx$ 2.6 %, Li $\approx$ 10^{-3} %, Ru $\approx$ 10^{-2} %, Cs $\approx$ 10^{-4} %); sea water contains, on average, around 2.9 % of NaCl and around 0.008 % of KCl and a quantity of $0.01 - 0.04$ g/l of rubidium and lithium. There are large deposits of NaCl and KCl, containing small percentages of LiCl, RbCl, and CsCl, originating from the evaporation of ancient inland seas. Sodium and potassium are also present in significant quantities in plants and other living beings.

Magnesium, a metal from the second group of the periodic table of elements, is also highly reactive and not to be found in nature in its elementary form. It constitutes about 2.5 % of the lithosphere. Bismuth, in the fifth group, constitutes about 2 10^{-5} % of the lithosphere and is mostly present in the form of sulphur compounds and oxides. Its melting temperature is excessively high (see Table 5.1). Mercury is the only metal in liquid phase at room temperature.

The reactor "clementine", of 25 kW, the first fast nuclear reactor, designed and built between 1945 and 1946 (reaching full power in 1949), used mercury as its coolant.

Sodium has been used as cooling fluid in fast reactors of the LMFBR (Liquid Metal Fast Breeder Reactors) type. Its high boiling point (see Table 5.1) means that a low pressure can be maintained within the nuclear section of the plant and, at the

Fig. 5.2 Vapour pressure as a
function of the temperature
for some alkali metals, for
mercury and for water

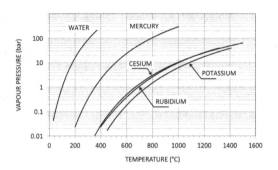

same time, obtain a steam with optimal thermodynamic characteristics (for example,
steam at 170 bar and 500 °C). The high heat exchange coefficients allow high power
densities (up to 800 kW/l). On the downside and a common characteristic of all
liquid metals, if the fluid gets contaminated by impurities (for instance, oxygen and
hydrogen), phenomena of steel corrosion can increase dramatically. Sodium, like all
alkali metals, reacts violently with oxygen in the air, and if any liquid (in the case of
minor leakage) is nebulised in the air, then the reaction is violent. In large quantities,
liquid sodium burns slowly on the exposed surface.

Other metals considered during the development phase of fast reactors include
potassium, caesium, lithium, and mixtures of sodium-potassium. For the reactors of
the so-called Generation IV, there are projects of fast reactors using lead or lead-
bismuth eutectic mixtures as coolant.

Excluding mercury, on account of its toxicity and limited production, of the
metals listed in Table 5.1, those with the highest vapour pressure are caesium,
rubidium, and potassium. Figure 5.2 shows the vapour pressure curves for caesium, rubidium, and potassium and, for comparison, mercury and water. Caesium
and rubidium have very similar vapour pressures; Caesium has a molar mass 1.55
times greater than rubidium, which implies smaller enthalpy drops in the turbine
and less evaporation heat (therefore, tending to a higher mass flow rate for the
same useful power), and higher vapor density at the same temperature and pressure.
Rubidium is much more abundant than caesium, in fact, similar to zinc. Potassium,
among the three alkali metals considered, has the lowest molar mass, is widely
available and at low cost, but, at the same temperature, has a lower vapour pressure
than rubidium: for example, at the temperature of 500 °C, the vapour pressure of
potassium is 0.042 bar, but that of rubidium is 0.11 bar. The useful power (at preset
dimensions and fixed condensation temperature) on a single exhaust flow turbine
therefore tends to be lower than that for caesium or rubidium.[4]

---

[4]The power $\dot{W}$ of the last stage of a turbine is, approximately, if calculating on a preset number
of Mach, like $\dot{W} = \dot{m}\Delta H \propto \rho D^2 v_s v_s^2 \propto (P/T) M D^2 v_s^3 \propto D^2 (P/T) M (T/M)^{3/2}$, with D
average diameter, ρ density of vapour, v_s speed of sound, ΔH enthalpy drop and M molecular
weight. For the same average diameter and the same condensation temperature, the power is,
therefore, approximately proportional to the ratio $P/\sqrt{M}$. For example, assuming $T = 500$ °C,

All the alkali metals are highly reactive, with the risk of fire, so air (and all compounds in general that contain oxygen, including water) must absolutely not be used in the circuits. In the case of potassium, for example, corrosion, too, increases dramatically in the presence of an oxygen concentration of just a few parts per million [2].

A broad discussion of the problems of corrosion for alkali metal, liquid, and vapour on structural materials and the problems of corrosion for combustion products can be found in [12]. Potassium at 870 °C seems less corrosive than steam in the presence of Ni − Cr alloys. The refractory metals do not manifest any corrosion with potassium up to 1,000 °C, but they cannot be used in combustion systems, given their high reactivity to oxygen. The maximum pressures in a cycle with alkali metals are hundred times lower than those in the USC steam cycles and the stress in the potassium tubes, for example, is very low: the austenitic steels seem adequate for potassium up to temperatures of 850–900 °C. One possibility for alleviating the problems of corrosion linked with the use of carbon with a high presence of sulphur could be the adoption of a pressurised fluidised bed combustor, in which a sorbent (limestone or dolomite), mixed with the carbon, traps the sulphur, reducing the corrosive action of the combustion gases [13, 14].

As far as potassium is concerned, 800,000 hours of testing have been accumulated, with operating temperatures even greater than 540 °C: in turbines with two or three stages at 815 °C, in boilers in the range 760–1,200 °C, in condensers at 600–870 °C, in mechanical and electromagnetic pumps and in throttle and shut-off valves [2]. In conclusion, a firm technological base exists for potassium Rankine systems, albeit for smaller components than those used in a large power station. However, a pilot plant has never been built.

Without doubt, planning and building systems with alkali liquid metals as the working fluids presents many difficulties. The characteristics of liquid metals, though, are unique (also in terms of conductibility and viscosity) and developing the technology in this sector could bring numerous benefits.

When a number N of thermodynamic cycles are superimposed, assuming that all the waste heat from a cycle is completely used and constitutes a unique heat source for the lower contiguous cycle, the efficiency η of the multiple cycle is a function of the efficiencies of the individual cycles that compose it, according to (see [12])

$$\eta = \eta_1 + \sum_{i=2}^{N} \eta_i \prod_{j=1}^{i-1} \left(1 - \eta_j\right) \tag{5.1}$$

with index 1 identifying the cycle with the highest temperature. In the case of a binary cycle, made by superimposing two cycles 1 and 2, (5.1) becomes

---

the power of the stage with rubidium would be approximately 1.7–1.8 times greater than that of the stage with potassium.

$$\eta = \eta_1 + \eta_2 (1 - \eta_1) \tag{5.2}$$

In the case of a ternary cycle,

$$\eta = \eta_1 + \eta_2 (1 - \eta_1) + \eta_3 (1 - \eta_1) (1 - \eta_2) \tag{5.3}$$

The various terms $(1 - \eta_j)$ take into account the fact that the thermal energy available for each cycle diminishes, passing from the upper to the lower thermodynamic cycle, because part of it is gradually converted into work.

5.2   The Binary Potassium and Rubidium–Steam Cycle

The most obvious solution in designing a binary liquid metal-steam cycle is that of condensing the metal vapours at a temperature just slightly higher than the maximum temperature of a steam cycle, according to the scheme in Fig. 5.3a. However, in the configuration shown in Fig. 5.3a, there is a great irreversibility of heat exchange between the metal vapours that condense and the vapour that preheats, evaporates, and superheats (an irreversibility due to the great difference between the condensation temperature of the metal cycle and the evaporation temperature of the steam). A more efficient solution from a thermodynamic point of view is shown in Fig. 5.3b, with two levels of condensation for the metal cycle. Multilevel condensation is thermodynamically advantageous, just like multilevel evaporation, which is typical of combined cycles with gas turbines and steam. Figure 5.3b, reports the total entropy and not the specific entropy for the expanding metal vapour: the entropy of the diagram is the sum of the entropy of the fluid that continues the expansion and the entropy of the vapour that condenses.

As a further example, a configuration like that of Fig. 5.4 (with the metal cycle at a maximum temperature equal to that of the steam cycle) would allow a significant reduction of the irreversibilities (that is, the thermodynamic imperfections) typical of the steam cycle: an average temperature of heat introduction originating from the primary energy source, which is far lower than the maximum temperature of the cycle. In Fig. 5.4, the condensation of the metal vapours at low temperature brings about the evaporation of the steam (from point 1 to point 2), and the condensation at a higher temperature is responsible for superheating the steam up to the temperatures (to be optimised) corresponding to points 3 and 5. Obviously, the final superheating and resuperheating must occur (in the scheme of Fig. 5.4) using thermal energy originating from the outside (sections from 3 to 6 and from 5 to 7) and (5.2) is not applicable. The overall efficiency is significantly increased, however. For instance (see [15]), it becomes 52 % with the steam cycle efficiency at 0.43 and the potassium topping cycle efficiency at 0.18 (maximum temperature 600 °C, minimum temperature 40 °C). In the examples illustrated in Figs. 5.3 and 5.4, the cycles shown are purely indicative.

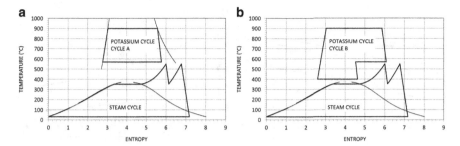

Fig. 5.3 Binary cycles with liquid metal and steam. (**a**) Metal cycle with single level of condensation, (**b**) Metal cycle with two levels of condensation

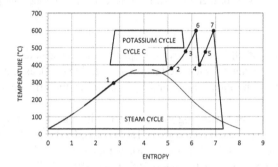

Fig. 5.4 Thermodynamic scheme of a topping cycle with liquid metal, with maximum temperature the same as the steam cycle

Referring to the drawings in Fig. 5.3, assuming a prefixed value (hereafter, 20 °C) of the pinch point temperature difference in the condenser of the liquid metal cycle and fixing a maximum temperature for the cycle with metal, the resulting binary cycle has an efficiency that depends on the maximum temperature of the steam cycle. In the case of a cycle with just one level of condensation (cycle A in Fig. 5.3a) Fig. 5.5a reports, for potassium and rubidium, the thermodynamic efficiency of the metal cycle as the maximum temperature varies, assuming a minimum temperature of 570 °C. The efficiencies are substantially the same (equal to about 0.19 when the maximum temperature is 850 °C). The main differences between the two metals lie in the different molar mass (39.1 for potassium, 85.48 for rubidium) and (having set the temperature) in the greater vapour pressure of rubidium with respect to potassium (for example, see Fig. 5.2, at 570 °C, 0.12 bar for potassium and 0.27 bar for rubidium). The greater molar mass reduces the work of expansion and the speed of sound, profoundly affecting the turbine design, and the greater vapour pressure, once the power is set, reduces the volume flows and the dimensions of the tubes and the turbine. Figure 5.5b compares (for cycle A) the metal vapour volumetric flow rates per MW of useful power of the metal cycle in the case of rubidium and potassium: the volume flows of potassium are almost double than those of rubidium.

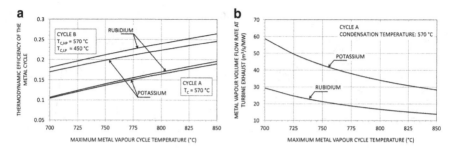

Fig. 5.5 (**a**) Efficiency of cycles with liquid metal with one and two levels of condensation as a function of maximum temperature. (**b**) Volume flows at the outlet of the turbine as maximum temperature varies per MW of useful power, in the case of a metal cycle with just one level of condensation

The efficiency of the binary cycle depends on the efficiency of the steam bottoming cycle and can be calculated using the relation (5.2), assuming η_1 is equal to the efficiency of the metal cycle and η_2 is equal to the efficiency of the steam cycle.

Assuming the steam bottoming cycle depicted in Figs. 1.25 and 1.29, with the basic assumptions for the calculations for Table 1.4 and a thermodynamic efficiency of 0.465 (see Exercise 1.7), we obtain, for a metal vapour cycle temperature of 850 °C (see Fig. 5.5a), an $\eta = 0.19 + 0.465(1 - 0.19) = 0.57$.

In general, once the maximum temperature of the metal cycle is set, there is a strict dependence of the efficiency η on the maximum temperature of the steam cycle. In the case of cycle A, an increase in the maximum temperature of the steam cycle reduces the η_1 efficiency and the efficiency η of the binary cycle tends to diminish with the increase in the maximum temperature of the steam cycle (even though, clearly, it will always be superior to the η_2 value of just the steam cycle). This occurs because of the rapid deterioration in the efficiency of the metal cycle as the condensation temperature rises, compared to the increase in efficiency of the steam cycle as the maximum temperature rises.

Figure 5.5a also represents the efficiency of potassium and rubidium cycles with two levels of condensation (at 570 °C and at 450 °C, cycle B of Fig. 5.3b), as the maximum temperature of the metal cycle varies. Condensation with two levels benefits significantly the thermodynamic performance of the cycle (for example, at the maximum temperature of 850 °C, in the case of potassium, the efficiency passes from 0.189 to 0.246). By contrast, the condensation pressures at the lower temperature diminish rapidly, with problems for the dimensions of the final stages of the turbine: in the case of potassium, for example, if the LP condensation temperature passes from 500 °C to 400 °C, the efficiency of the metal cycle passes from 0.225 to 0.261 (with a maximum temperature of 850 °C and an HP condensation temperature of 570 °C), but the volume flow at the outlet of the power unit increases from 280.8 m³/s/MW to 612.6 m³/s/MW.

In the case of cycle B of Fig. 5.3b, as the maximum temperature of the steam cycle rises (once the maximum temperature of the metal cycle is set), unlike the case with cycle A, there exists an optimal value for the efficiency of the binary cycle, although not particularly pronounced; see [12].

Topping Cycle Turbine Design

In principle, designing turbines with metal vapours is not much more complicated than for the large steam turbines. The overall pressure ratio of the steam cycle in Fig. 1.29, for example, is equal to 4,000, for a cycle with potassium or rubidium, between 850 °C and 400 °C, though, it is 372 and 161, respectively. The specific work of the metal vapours is significantly inferior to that of steam: a typical cycle among those considered produces 500–250 kJ/kg and 200–100 kJ/kg for potassium and rubidium, respectively. These compare with values of 1,000–1,500 kJ/kg in the case of typical steam cycles.

Only with regard to vapour quality during expansion are the cycles with metal vapours critical (typically, for the cases we have considered, the vapour quality is 0.74–0.77, at the end of expansion). A similar situation can be found in traditional nuclear steam cycles, and the problem can be tackled by resorting to adequate fluid dynamics design and to extraction devices similar to those adopted in steam cycles [16–18].

As an example of the dimensions in designing a turbine with metal vapours, with reference to cycle B of Fig. 5.3b, having set an acceptable maximum dimension for the final stage,[5] it is possible, for each working fluid, having identified the condensation temperature $T_{C,LP}$, to evaluate the corresponding vapour flows. An eight-stage configuration has been assumed for the turbine (with two LP flows and three stages for each flow and five stages for the expansion at high pressure) in such a way that all the stages are subsonic. As another calculation hypothesis, the kinetic energy at the outlet of each stage is kept at 3–4 % of the enthalpy drop.

The temperature $T_{C,HP}$ is fixed at 570 °C, the $T_{C,LP}$ at 450 °C, the maximum temperature $T_E = 850$ °C and the bottoming steam cycle of reference is that in Figs. 1.25 and 1.29 (see also Exercise 1.7).

Several results for the turbine with potassium (the fluid at the lowest condensation pressures, having fixed the temperatures $T_{C,HP}$ and $T_{C,LP}$), and rubidium as working fluids are shown in Tables 5.2 and 5.3. For the turbine with potassium, the number of revs is set at 1,500 rpm, and the tip diameter of the final stage at low pressure is 5.4 m (see Table 5.2), with a h/D ratio of 0.277. The volume flows between the first and last stage vary between 74 and 1,461 m^3/s.

---

[5]In the case of steam turbines, stages have been designed with a tip diameter of 4.32 m at 3,000 rpm and stages with a tip diameter of 6.7 m at 1,500 rpm with h/D ratios, height of blade with respect to average diameter, of 0.39 and 0.37, respectively.

Table 5.2 Some characteristics of the double-flow LP turbine for potassium and rubidium

LP turbine — Two flows — Three stages		Potassium	Rubidium
Number of revolution		1, 500 rpm	750 rpm
Overall pressure ratio		7.14	5.14
Overall isentropic work		307 kJ/kg	75 kJ/kg
Isentropic power per stage		1, 843 kW	2, 116 kW
First stage	Tip diameter	4.38 m	4.38 m
	Mean diameter	4.01 m	3.95 m
	h/D [a]	0.0926	0.11
	Reaction degree at mean radius	0.5	0.5
	Volume flow ratio	1.67	1.49
	Exhaust volume flow	446 m^3/s	247 m^3/s
Second stage	Tip diameter	4.65 m	4.62 m
	Mean diameter	4.01 m	3.95 m
	h/D [a]	0.161	0.173
	Reaction degree at mean radius	0.5	0.5
	Volume flow ratio	1.76	1.59
	Exhaust volume flow	784 m^3/s	392 m^3/s
Third stage	Tip diameter	5.40 m	5.27 m
	Mean diameter	4.22 m	4.16 m
	h/D [a]	0.277	0.266
	Reaction degree at mean radius	0.55	0.55
	Volume flow ratio	1.87	1.66
	Exhaust volume flow	1461 m^3/s	651 m^3/s

[a] Mean blade high to mean diameter ratio

In the case of rubidium, the LP section is chosen with geometric dimensions that are approximately the same as those of the turbine with potassium and the 750 rpm represents the closest synchronous velocity to that which guarantees a specific number ω_S at each stage, close to those of the stages with potassium.

The tip peripheral velocities of the turbine stages with rubidium are around half those of potassium: 125–207 m/s, from the first to the last stage, compared to 240–424 m/s. The turbine power with potassium is 36.8 MW and with rubidium is 43.6 MW.

Some Concluding Remarks

Analysis and discussion have shown how there already exist materials that are capable of enabling the use of alkali metals in power cycles superimposed on standard steam cycles, with global thermodynamic efficiencies of 55–60 % and relatively modest maximum temperatures (750–850 °C) and evaporation pressures for the metal section of just a few bar (3.8 bar, for rubidium at 850 °C). The specific

Table 5.3 Some characteristics of the HP section of turbines with potassium and rubidium

HP turbine — Single flow — Five stages		Potassium	Rubidium
Number of revolution		1, 500 rpm	750 rpm
Overall pressure ratio		18.6	14.1
Overall isentropic work		519 kJ/kg	134 kJ/kg
Isentropic power per stage		5, 148 kW	6, 172 kW
First stage	Tip diameter	3.02 m	3.17 m
	Mean diameter	2.93 m	3.05 m
	h/D [a]	0.029	0.04
	Reaction degree at mean radius	0.06	0.1
	Volume flow ratio	1.5	1.4
	Exhaust volume flow	73.8 m^3/s	54.8 m^3/s
Last stage	Tip diameter	4.14 m	4.39 m
	Mean diameter	3.52 m	3.68 m
	h/D [a]	0.176	0.192
	Reaction degree at mean radius	0.35	0.38
	Volume flow ratio	1.82	1.7
	Exhaust volume flow	582 m^3/s	344 m^3/s

[a] Mean blade high to mean diameter ratio

isentropic work, far inferior to that of steam, means that just a few turbine stages need to be used and these are not particularly stressed from a mechanical point of view.

In the ordinary steam cycles, a rise in the maximum temperature up to 700–800 °C would not raise efficiency to those levels obtained with a binary cycle, due to the fact that the average temperature of heat introduction is significantly lower than the maximum temperature.

On the other hand, condensation pressures in the metal cycle tend to be very low, which leads to very large volume flows at turbine exhaust, with serious fluid dynamic problems that limit the maximum power obtainable, making it very difficult, in practice, to build large-scale plants

In any case, even though this particular technology has interesting thermodynamic possibilities, there is no specific research activity taking place at present in the sector of binary cycles with liquid metals.

References

1. DiPippo R (2008) Geothermal power plants: principles, applications, case Studies, and environmental impact, 2nd edn. Butterworth-Heinemann, Oxford UK
2. Gutstein M, Furman ER, Kaplan GM (1975) Liquid-metal binary cycles for stationary power. NASA Technical Note, NASA TN D-7955

3. Gaffert GA (1952) Steam power stations, 4th edn. McGraw-Hill and Kōgakusha Company, New York and Tokyo

4. Fraas AP (October 1966) A potassium-steam binary vapor cycle for a molten-salt reactor power plant. J Eng Power Trans ASME, 355–366

5. Goldstein S, Vrillon B (1982) Confirmation of the advantages of a thermodynamic cycle using sulfur with the help of an experimental facility. In: Energy conservation in industry: combustion, heat recovery, and Rankine cycle machines. Proceedings of the contractors' meetings, Brussels. D.Reidel Publishing Company, Dordrecht, pp. 213–222

6. Fraas AP (1973) A potassium-steam binary vapor cycle for better fuel economy and reduced thermal pollution. J Eng Power Trans ASME 95(1):53–63

7. Dow HH (1926) Diphenyl oxide bi-fluid power plants. J Am Soc Nav Eng 38(4):940–950

8. Killeffer DH (1935) Stable organic compounds in power generation. Diphenyl-Diphenyl oxide mixtures in an efficient boiler plant of unique design. Ind Eng Chem 27(1):10–15

9. d'Amelio L (1936) The steam turbine and the binary cycles with fluids other than water between the lower isotherms. L'elettrotecnica 23(9):250–257 (in Italian)

10. Fleury J, Bellot Ch (1984) Ammonia bottoming cycle development at Electricité de France for nuclear power plants. In: ORC-HP-Technology. Working Fluid Problems. Proceedings of the International VDI-Seminar, Zürich. VDI-Verlag GmbH, Düsseldorf, pp. 221–241

11. Angelino G, Invernizzi C, Molteni G (1999) The potential role of organic bottoming Rankine cycles in steam power stations. Proc IME J Power Energ 213:75–81

12. Angelino G, Invernizzi C (2006) Binary and ternary liquid metal-steam cycles for high-efficiency coal power stations. Proc IME J Power Energ 220:195–205

13. Fraas AP, Brooks RD (1974) Topping and bottoming cycles. Paper prepared for presentation at the 9th World Energy Conference, Detroit, Michigan, 22–27 September

14. Fraas AP (1976) Application of the fluidized bed coal combustion system to the production of electric power and process heat. Paper prepared for presentation at the American Institute of Chemical Engineers. Kansas City, Missouri, 11–14 April

15. Angelino G, Invernizzi C (2008) Binary conversion cycles for concentrating solar power technology. Sol Energ 82:637–647

16. Manson SV (1968) A review of the alkali metal Rankine technology program. J Spacecraft Rockets 5(11):1249–1259

17. Fraas AP, Burton DW, LaVerne ME, Wilson LV (1969) Design comparison of cesium and potassium vapor turbine-generator units for space power plants. Oak Ridge National Laboratory, Oak Ridge, Tennessee. ORNL-TM-2024

18. Fraas AP (1975) A cesium vapor cycle for an advanced LMFBR. Paper submitted for presentation at the ASME Winter Annual Meeting. Houston, Texas, 30 November–5 December

Appendix A
The Fluid Machines and the Balance Equations

A fluid machine, or generically a fluid system, is a system in which one or more fluids perform an energy conversion by means of dynamical and kinematical processes. The operating fluids are generally named energy carriers of the transformation process [1]. Figure A.1 represents a fluid apparatus consisting of two fluid machines (a heat exchanger and a turbomachine) and, in addition, some ducts to convey the fluids. The solid walls of the system usually delimit a fixed volume in the space and one or more energy carriers pass through it. In Fig. A.1, $\mathbf{n}_1$ and $\mathbf{n}_2$ represent the unit vectors in the direction of the fluid in question (1 at the entrance and 2 at the exit to the system). Let us assume A_1 and A_2 are the cross sections normal to the fluid directions $\mathbf{n}_1$ and $\mathbf{n}_2$ and $\mathbf{v}_1 = v_1 \mathbf{n}_1$ and $\mathbf{v}_2 = v_2 \mathbf{n}_2$ the mean fluid velocities on cross sections 1 and 2. The powers $\dot{Q}$ and $\dot{W}$ in Fig. A.1 represent, respectively, the thermal power and the mechanical power supplied by the environment to the fluid system concerned.

The balance equations are drawn from, and presented in, numerous textbooks. A variety of different and interesting points of view and ways of presentation can be found in [2–7]. The balance equations which are presented and discussed below are usually applied to the fluid passing through the system. The system could also be closed at the mass: in this case, there is no mass flow over cross sections 1 and 2, and the fluid remains confined within the system we are considering.

A.1 The Mass Balance

According to the following assumptions,

1. The mean velocities $\mathbf{v}_1$ and $\mathbf{v}_2$ are normal to sections A_1 and A_2.
2. The densities ρ_1 and ρ_2 of the fluid are uniform over sections A_1 and A_2

the mass balance gives

$$\frac{\mathrm{d}}{\mathrm{d}t} M_{\text{tot}} = \rho_1 v_1 A_1 - \rho_2 v_2 A_2 \tag{A.1}$$

C.M. Invernizzi, *Closed Power Cycles*, Lecture Notes in Energy 11, DOI 10.1007/978-1-4471-5140-1, © Springer-Verlag London 2013

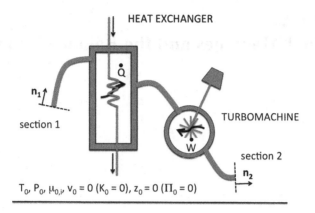

Fig. A.1 A schematic representation of a fluid system. K_0 and Π_0 are, respectively, the kinetic energy and potential energy associated with velocity v_0 and quota z_0

with

$$M_{\text{tot}} \equiv \int_{\mathcal{V}} \rho \, d\mathcal{V} \quad \text{that is, the total mass of the fluid in the considered system}$$

and with

$$\frac{d}{dt} M_{\text{tot}} \equiv \quad \text{the time variation (the rate of increase) of the total mass in the system}$$

$\rho_1 v_1 A_1 \equiv \quad$ the mass flow on the section 1, $\quad \dot{m}_1$

$\rho_2 v_2 A_2 \equiv \quad$ the mass flow on the section 2, $\quad \dot{m}_2$

Each individual term of (A.1) is expressed in kg/s. If $dM_{\text{tot}}/dt = 0$, the system is said to be in stationary conditions and $\dot{m}_1 = \rho_1 v_1 A_1 = \dot{m}_2 = \rho_2 v_2 A_2$. If, furthermore, $\rho_1 = \rho_2$ (incompressible fluid), $v_1 A_1 = v_2 A_2$.

Exercises

A.1. With reference to a piston pump in periodic motion in a cylindrical chamber (see Fig. A.2), let us calculate the delivery mass flow as a function of time in the case of an ideal pump (volumetric efficiency equals to one).

From Fig. A.2, setting $\lambda = r/L$ and $c = 2r$, the following geometrical relations are valid:

$$L \sin \beta = r \sin (\pi - \alpha)$$

$$L \sin \beta = r \sin \alpha$$

Fig. A.2 Schematic representation of a volumetric piston pump. Geometrical parameters and reference system

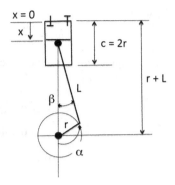

$$\sin \beta = \lambda \sin \alpha$$

and

$$\cos \beta = \sqrt{1 - \sin^2 \beta} = \sqrt{1 - \lambda^2 \sin^2 \alpha}$$

Thus, the coordinate x which represents the position of the piston with respect to the upper dead point based on the crank angle α is

$$x = (r + L) - (L \cos \beta + r \cos (\pi - \alpha))$$
$$= (r + L) - (L \cos \beta - r \cos \alpha)$$
$$= r (1 + \cos \alpha) + L \left(1 - \sqrt{1 - \lambda^2 \sin^2 \alpha}\right)$$

When

$$\alpha = 0 \qquad \text{the piston is at the lower dead point} \quad (x = 2r)$$
$$\alpha = \pi \qquad \text{the piston is at the upper dead point} \quad (x = 0)$$

We can now evaluate the velocity of the piston

$$v = \frac{dx}{dt}$$
$$= \frac{d\alpha}{dt} \frac{dx}{d\alpha}$$
$$= \omega \left(-r \sin \alpha + \frac{r}{\lambda} \frac{1}{2} \frac{\lambda^2 2 \sin \alpha \cos \alpha}{\sqrt{1 - \lambda^2 \sin^2 \alpha}}\right)$$
$$= \omega \left(-r \sin \alpha + \frac{L \lambda^2 \sin \alpha \cos \alpha}{\sqrt{1 - \lambda^2 \sin^2 \alpha}}\right)$$

with

$$\omega = \frac{d\alpha}{dt} = \frac{2\pi N}{60} \qquad \text{the angular velocity}$$

and N representing the round per minute.

The velocity is

$$v = 0 \quad \text{when} \quad \alpha = 0 \quad \text{and} \quad \alpha = \pi$$

$$v < 0 \quad \text{when} \quad 0 < \alpha < \pi$$

$$v > 0 \quad \text{when} \quad \pi < \alpha < 2\pi$$

Finally, the delivery mass flow can be evaluated by (A.1)

$$\frac{dM_{\text{tot}}}{dt} = -\dot{m}_2$$

As

$$M_{\text{tot}} = A\rho x\,(t)$$

then

$$\frac{dM_{\text{tot}}}{dt} = A\rho \frac{dx}{dt}$$

and

$$\dot{m}_2 = -A\rho \frac{dx}{dt} = -A\rho v$$

$$= A\rho\omega \left(r\sin\alpha - \frac{L\lambda^2 \sin\alpha \cos\alpha}{\sqrt{1 - \lambda^2 \sin^2 \alpha}} \right) \tag{A.2}$$

The delivery mass flow is

$$\dot{m}_2 = 0 \quad \text{when} \quad \alpha = 0 \quad \text{at the lower dead point}$$

$$\dot{m}_2 = 0 \quad \text{when} \quad \alpha = \pi \quad \text{at the upper dead point}$$

$$\dot{m}_2 = A\rho\omega r \quad \text{when} \quad \alpha = \frac{\pi}{2}$$

The flow $\dot{m}_2$ varies from a minimum equal to zero (when the piston is in the lower dead point and in the upper dead point) to its maximum value, corresponding to $\alpha = \pi/2$. The maximum value of the flow mass rises linearly with the number of revs.

A.2. Referring to Fig. A.2, we calculate the mean flow $\overline{m}$ on the cycle, using (A.2), with reference to an ideal machine (volumetric efficiency equals to one).

The mean flow is calculated with reference to the period, but the useful flow phase is only over the semi-period $0 - \pi$. So

$$\overline{m}_2 = \frac{1}{2\pi} \int_0^\pi \dot{m}_2 d\alpha$$

$$= \frac{1}{2\pi} \int_0^\pi A\rho\omega \left(r \sin\alpha - \frac{L\lambda^2 \sin\alpha \cos\alpha}{\sqrt{1 - \lambda^2 \sin^2\alpha}} \right) d\alpha$$

$$= \frac{A\rho\omega}{2\pi} \left[-r\cos\alpha + \sqrt{1 - \lambda^2 \sin^2\alpha} \right]_0^\pi$$

$$= A(2r)\rho\frac{N}{60}$$

with N as the number of revs/min. The mean flow over the cycle, irrespectively of the delivery pressure P_2, increases with the displacement $A \times 2r = A \times c$ and the number of revs N.

A.2 The Energy Balance

Assuming, with reference to Fig. A.1, that

1. The mean fluid velocities v_1 and v_2 on cross sections 1 and 2 are normal to A_1 and A_2
2. All the physical properties of the fluid are uniform on the inlet and on the outlet sections 1 and 2
3. The viscous forces on sections 1 and 2 are negligible so that the work of the viscous forces (negligible) is smaller than the pressure work
4. The thermal energy transported by conduction through A_1 and A_2 is negligible in comparison to the convective term
5. It is possible to define locally all the required thermodynamic equilibrium properties of the fluid (P, T, U, etc.)

the conservation of energy can be written as

$$\frac{d}{dt}(U_{tot} + K_{tot} + \Pi_{tot}) = \left(\rho_1 U_1 v_1 + \frac{1}{2}\rho_1 v_1^3 + \rho_1 \Pi_1 v_1 \right) A_1$$

$$- \left(\rho_2 U_2 v_2 + \frac{1}{2}\rho_2 v_2^3 + \rho_2 \Pi_2 v_2 \right) A_2$$

$$+ \dot{Q} + \dot{W} + P_1 v_1 A_1 - P_2 v_2 A_2 \qquad (A.3)$$

in which

$$U_{tot} \equiv \int_{\mathcal{V}} \rho U d\mathcal{V} \quad \text{is the total internal energy of the fluid contained in the system}$$

and U (the specific internal energy)

$K_{\text{tot}} \equiv \displaystyle\int_{\mathscr{V}} \frac{1}{2}\rho v^2 \mathrm{d}\mathscr{V}$ is the global kinetic energy

$\Pi_{\text{tot}} \equiv \displaystyle\int_{\mathscr{V}} \rho\Pi \mathrm{d}\mathscr{V}$ is the global potential energy

Then,

$\dfrac{\mathrm{d}}{\mathrm{d}t}(U_{\text{tot}} + K_{\text{tot}} + \Pi_{\text{tot}}) \equiv$ the time variation of the total energy

$\left(\rho_1 U_1 v_1 + \dfrac{1}{2}\rho_1 v_1^3 + \rho_1 \Pi_1 v_1\right) A_1 \equiv$ the energy flow across the section 1

$\left(\rho_2 U_2 v_2 + \dfrac{1}{2}\rho_2 v_2^3 + \rho_2 \Pi_2 v_2\right) A_2 \equiv$ the energy flow across the section 2

and

$\dot{Q} \equiv$ thermal power from the surrounding to the fluid in the system

$\dot{W} \equiv$ mechanical power from the surrounding to the fluid

$P_1 v_1 A_1 - P_2 v_2 A_2 \equiv$ power exchanged with the fluid across the sections 1 and 2

Introducing the mass flows $\dot{m}_1 = \rho_1 v_1 A_1$ and $\dot{m}_2 = \rho_2 v_2 A_2$ in (A.3) and defining the total energy $E_{\text{tot}} = U_{\text{tot}} + K_{\text{tot}} + \Pi_{\text{tot}}$, we obtain

$$
\begin{aligned}
\frac{\mathrm{d}}{\mathrm{d}t}E_{\text{tot}} = {} & \left(U_1 + \frac{P_1}{\rho_1} + \frac{1}{2}v_1^2 + \Pi_1\right)\dot{m}_1 \\
& -\left(U_2 + \frac{P_2}{\rho_2} + \frac{1}{2}v_2^2 + \Pi_2\right)\dot{m}_2 \\
& + \dot{Q} + \dot{W}
\end{aligned}
\tag{A.4}
$$

In the stationary conditions, E_{tot} is constant and $\dot{m}_1 = \dot{m}_2 = \dot{m}$. Then, in the gravitational field[1], $\Pi_1 = gz_1$ and $\Pi_2 = gz_2$. Furthermore, we can define the "static" enthalpy as $H = U + P/\rho$ and transform (A.4) into the new equivalent equation.

---

[1] The gravitational force, responsible for the force weight $\mathbf{F} = \rho\mathbf{g}$ per unit of volume, is conservative. It derives from a potential and admits a potential energy Π so that $\mathbf{F} = -\rho\nabla\Pi$. That is, $\Pi = gz + \Pi_0$. The main forces intervening in nature derive from a potential (apart from the forces of gravitation, there are electrostatic and magnetostatic ones).

$$\left(H_1 + \frac{1}{2}v_1^2 + gz_1\right)\dot{m} - \left(H_2 + \frac{1}{2}v_2^2 + gz_2\right)\dot{m} + \dot{Q} + \dot{W} = 0$$

that is,

$$\left(H_1 + \frac{1}{2}v_1^2 + gz_1\right) - \left(H_2 + \frac{1}{2}v_2^2 + gz_2\right) + Q + W = 0 \qquad \text{(A.5)}$$

In (A.4) each term represents a power (W). Each term of (A.5) represents, meanwhile, an energy specific to the mass (J/kg). We now apply the energy balance to some simple thermal machines.

Exercises

A.3. Referring to Fig. A.2, we calculate the power and the work to be supplied to the volumetric pump during the discharge phase.

To calculate the power, we apply (A.3) to the system in Fig. A.2 during the delivery of the mass flow. We ignore the contributions of the kinetic energy K_{tot} and the potential energy Π_{tot}, the kinetic energy of the fluid at the outlet $\frac{1}{2}\rho v_2^2$ and its quota of potential energy Π_2. We assume that compression is adiabatic ($\dot{Q} = 0$) and the fluid is incompressible ($\rho = const$). Thus, (A.3) becomes

$$\frac{\mathrm{d}}{\mathrm{d}t}(U_{tot}) = -\rho_2 U_2 v_2 A_2$$
$$+ \dot{W} - P_2 v_2 A_2$$

If the machine is ideal, the specific internal energy (a function only of the temperature for an incompressible fluid) remains constant. In this way, using the mass balance,

$$\frac{\mathrm{d}}{\mathrm{d}t}(U_{tot}) = U\frac{\mathrm{d}M_{tot}}{\mathrm{d}t} = U(-\dot{m}_2)$$

and therefore

$$U(-\dot{m}_2) = -\rho_2 U_2 v_2 A_2$$
$$+ \dot{W} - P_2 v_2 A_2$$

Since, on the other hand, $\rho_2 v_2 A_2 = \dot{m}_2$, in conclusion, we get

$$U(-\dot{m}_2) = U(-\dot{m}_2)$$
$$+ \dot{W} - P_2 v_2 A_2$$

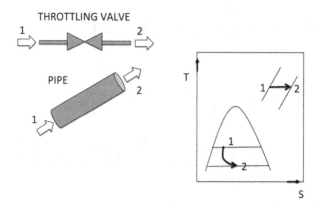

Fig. A.3 Throttling processes of a fluid. Representation of the thermodynamic process in the T-S plane

or

$$\dot{W} = P_2 v_2 A_2 = \frac{P_2}{\rho} \dot{m}_2 = P_2 A \omega \left(r \sin \alpha - \frac{L\lambda^2 \sin \alpha \cos \alpha}{\sqrt{1 - \lambda^2 \sin^2 \alpha}} \right)$$

which supplies the power as a function of the α angle during the delivery phase.

The work W is calculated by integrating the preceding equation on the duration of the delivery phase:

$$
\begin{aligned}
W &= \int_0^\pi \dot{W} \frac{dt}{d\alpha} d\alpha = \int_0^\pi \frac{\dot{W}}{\omega} d\alpha \\
&= P_2 A \int_0^\pi \left(r \sin \alpha - \frac{L\lambda^2 \sin \alpha \cos \alpha}{\sqrt{1 - \lambda^2 \sin^2 \alpha}} \right) d\alpha \\
&= P_2 A \left[-r \cos \alpha + \sqrt{1 - \lambda^2 \sin^2 \alpha} \right]_0^\pi \\
&= P_2 A (2r)
\end{aligned}
$$

with $A(2r)$ equal to the volume of the cylinder (Fig. A.2). The work during the delivery phase is directly proportional to the delivery pressure P_2 and the displacement. In a similar way, we can calculate the work during the aspiration phase and then the work over the whole cycle, the sum of the two works (that of the delivery and the aspiration), obtaining $(P_2 - P_1)(A2r)$.

A.4. We consider a throttling valve; see Fig. A.3. The fluid crossing the valve undergoes an adiabatic pressure reduction without work extraction:

$$Q = W = 0$$

$$g z_1 \approx g z_2$$

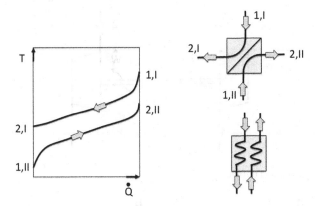

Fig. A.4 Schemes of surface heat exchangers and heat transfer diagram T-$\dot{Q}$

$$\frac{1}{2}v_1^2 \approx \frac{1}{2}v_2^2 \qquad \text{so, as a consequence,}$$

$$H_1 = H_2 \qquad \text{isoenthalpic expansion}$$

If the fluid is an ideal gas, $T_1 = T_2$.

A.5. A surface heat exchanger performs a transfer of thermal energy (heat) between two streams of fluid, each one at constant pressure and without extraction of work (see Fig. A.4). Usually, for these thermal machines, $\Delta \left(1/2v^2\right)$ and Δz are smaller than ΔH.

With index I we indicate the hot fluid and we assign index II to the cold fluid:

$$\dot{m}_I \left(H_{1,I} - H_{2,I}\right) = -\dot{Q}_I \qquad H_{1,I} > H_{2,I}$$

$$\dot{m}_{II} \left(H_{1,II} - H_{2,II}\right) = -\dot{Q}_{II} \qquad H_{1,II} < H_{2,II}$$

$$\text{with} \quad |\dot{Q}_I| = |\dot{Q}_{II}|$$

In the general case, $\dot{m}_I \neq \dot{m}_{II}$ and the two fluids can be of different nature too.

A.6. In the case of an adiabatic compressor (Fig. A.5),

$$\dot{Q} = 0$$

$$\dot{m}_1 = \dot{m}_2 = \dot{m}$$

$$\frac{1}{2}v_1^2 \approx \frac{1}{2}v_2^2$$

$$gz_1 \approx gz_2 \qquad \text{then}$$

$$(H_1 - H_2) + W = 0$$

$$W = H_2 - H_1$$

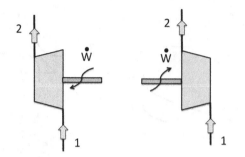

Fig. A.5 Compressor and turbine

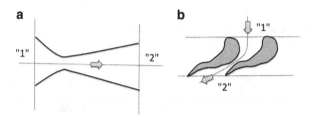

Fig. A.6 Examples of adiabatic nozzles

According to the assumed hypotheses, the specific work W is positive: from the surroundings to the fluid flowing through the compressor.

For an adiabatic turbine (Fig. A.5),

$$\dot{Q} = 0$$

$$\dot{m}_1 = \dot{m}_2 \quad = \quad \dot{m}$$

$$\frac{1}{2}v_1^2 \approx \frac{1}{2}v_2^2$$

$$gz_1 \approx gz_2 \quad \text{then}$$

$$(H_1 - H_2) + W = 0$$

$$W = -(H_1 - H_2)$$

According to our assumed hypotheses, the specific work W is negative: from the working fluid to the surroundings.

A.7. We consider now an adiabatic nozzle. The adiabatic nozzle is a device whose main purpose is transforming energy of a thermodynamic nature into kinetic energy. It is an adiabatic system which produces no useful work (Fig. A.6). Therefore,

$$\dot{Q} = W = 0$$

$$\dot{m}_1 = \dot{m}_2 = \dot{m}$$

$$gz_1 \approx gz_2$$

Therefore,

$$\left(H_1 + \frac{1}{2}v_1^2\right) - \left(H_2 + \frac{1}{2}v_2^2\right) = 0 \qquad \text{and}$$

$$H_1 - H_2 = \frac{1}{2}\left(v_2^2 - v_1^2\right)$$

The decrease in the static enthalpy changes into an increase of the specific kinetic energy (the fluid velocity).

A.8. With reference to an adiabatic and isentropic expansion of a perfect gas in a nozzle, let us calculate the fluid velocity on various sections along the symmetry axis (see Fig. A.6).

The thermodynamic conditions on the inlet section 1 and on the discharge section 2 are $P_1 = 10$ bar, $T_1 = 1,000\ °C$ (1273.15 K) and $P_2 = 1$ bar.

For a perfect gas,

$$\frac{p}{\rho} = \frac{R}{M}T \quad \text{(the volumetric equation of state)}$$

$$C_p = \text{constant}$$

$$C_p - C_v = \frac{R}{M}$$

$$\gamma = \frac{C_p}{C_v} \quad \text{(constant)}$$

$$H = H_0 + C_p\,(T - T_0)$$

In stationary conditions, with $z_1 = z_2$ and $v_1 \approx 0$,

$$\left(H_1 + \frac{1}{2}v_1^2\right) - \left(H_2 + \frac{1}{2}v_2^2\right) = 0$$

$$\frac{1}{2}\left(v_2^2 - v_1^2\right) = (H_1 - H_2)$$

$$\frac{1}{2}v_2^2 \approx H_1 - H_2$$

$$= C_p\,(T_1 - T_2)$$

$$= C_p T_1\left(1 - \frac{T_2}{T_1}\right)$$

Table A.1 Isentropic expansion of dry air in an adiabatic nozzle

p, bar	10	9	8	7	6	5	4	3	2	1
v, m/s	0	193	397	489	589	678	768	863	971	1110.4
T, °C	1,000	962	921	878	827	771	707	629	531	386.3
A^*, cm^2/kg/s	∞	14.30	10.78	9.47	8.93	8.84	9.16	10.1	11.88	17.04
A^*/A^*_{throat}	∞	1.62	1.22	1.07	1.01	1.0	1.04	1.14	1.33	1.93
M	0	0.39	0.57	0.73	0.89	1.05	1.22	1.43	1.71	2.16

For an isentropic expansion $p/\rho^\gamma = $ constant, so resorting to the volumetric equation of state,

$$\frac{P_1}{\left(\frac{P_1 M}{RT_1}\right)^\gamma} = \frac{P_2}{\left(\frac{P_2 M}{RT_2}\right)^\gamma}$$

$$\left(\frac{P_1}{P_2}\right)^{1-\gamma} = \left(\frac{T_2}{T_1}\right)^\gamma$$

$$\frac{T_2}{T_1} = \left(\frac{P_1}{P_2}\right)^{\frac{1-\gamma}{\gamma}} = \left(\frac{P_2}{P_1}\right)^{\frac{\gamma-1}{\gamma}}$$

then

$$\frac{1}{2}v_2^2 = C_p T_1 \left(1 - \left(\frac{P_2}{P_1}\right)^{\frac{\gamma-1}{\gamma}}\right)$$

as $C_p = \gamma R/M (\gamma - 1)$, we obtain

$$\frac{1}{2}v_2^2 = \frac{\gamma R}{M (\gamma - 1)} T_1 \left(1 - \left(\frac{P_2}{P_1}\right)^{\frac{\gamma-1}{\gamma}}\right)$$

$$v_2 = \sqrt{2\frac{\gamma R}{M (\gamma - 1)} T_1 \left(1 - \left(\frac{P_2}{P_1}\right)^{\frac{\gamma-1}{\gamma}}\right)}$$

with $R = 8.3143$ J/mol K and M is the molar mass of the gas. For dry air, M = 28.97 kg/kmol, $C_p = 1.0035$ kJ/kg K, $C_v = 0.7165$ kJ/kg K and $\gamma = 1.4$

Assuming a unit mass flow $\dot{m}$ it is possible now to evaluate the gas velocity as a function of the pressure, from 10 bar to 1 bar. The results are in Table A.1.

The quantity A^* ($= 1/\rho v$) is the area section for a unit mass flow and M is the Mach number ($= v/c$, with $c = \sqrt{(\partial P/\partial \rho)_S}$) the speed of sound).[2] Section A^*

---

[2]For an ideal gas, $c = \sqrt{\gamma (R/M) T}$. For air at 25 °C and 1 bar, $c = 346$ m/s. For water at 25 °C (practically, an incompressible fluid), $c = 1497$ m/s.

decreases during the first part of the expansion process, up to its minimum value A^*_{throat}. Then, during the supersonic expansion, it increases.

A.3 The Energy Balance with Chemical Reactions

Usually, the definition of a "reference state" for the state variables internal energy U, enthalpy H, and entropy S is not a problem because they appear in the energy balances as simple differences and the knowledge of their "absolute" value before and after the process is unimportant.

Different are the cases in which the energy carriers involved in the transformation process undergo chemical reactions: in these cases, the reactants are physically different from the products, and in the equation of energy balance, the formation enthalpies must be added. The standard formation enthalpy $\Delta_f H_i^0$ is the enthalpy required to form one mole of the species "i" in its standard state conditions (room temperature T_0 and atmospheric pressure P_0).[3] The values of $\Delta_f H_i^0$ for every single chemical species are measured and tabulated.

Exercises

A.9. Let's calculate the enthalpy of reaction of the combustion of methane at T_0 and P_0 in a reactor operating at constant pressure (see Fig. A.7).

The reference chemical reaction is

$$CH_{4(g)} + 2O_{2(g)} \rightarrow CO_{2(g)} + 2H_2O_{(g)}$$

Reactants and products are at the standard temperature T_0 (25 °C) and pressure P_0 (one metric atmosphere, 0.98 bar). The thermal power $\dot{Q}$ per unit fuel mass flow represents the standard heat (enthalpy) of reaction $\Delta_r H^0$. If, among the products, water is in the gaseous state, the standard heat of reaction is named lower heating value (LHV); if water is in the liquid state, $\Delta_r H^0$ represents the higher heating value (HHV).

Because, usually, the enthalpies of formation $\Delta_f H_i^0$ are given with reference to the mole, the energy balance has to be written in molar terms

$$\dot{Q} = \Delta_r H^0 = \sum_i \dot{n}_i \Delta_f H^0_{\text{products},i} - \sum_j \dot{n}_j \Delta_f H^0_{\text{reactants},j}$$

---

[3] The standard state pressure P_0 is 1 atm = 0.98 bar. The standard room temperature T_0 is 298.15 K (25 °C).

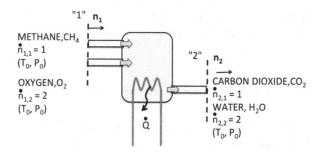

Fig. A.7 Scheme of a combustor at constant pressure. The fuel is methane and oxygen is the second reactant. Carbon dioxide and water are the products of the combustion

In the specific case [8], $\Delta_f H^0_{CO_2(g)} = -393.52$ kJ/mol, $\Delta_f H^0_{H_2O(g)} = -241.83$ kJ/mol, $\Delta_f H^0_{CH_4(g)} = -74.87$ kJ/mol, $\Delta_f H^0_{O_2(g)} = 0.0$ kJ/mol, then

$$\Delta_r H^0 = [-393.52 + 2\,(-241.83)] - [-74.87 + 2\,(0.0)] = -802.31 \text{ kJ/mol}$$

As the molar mass of methane is 16.034 g/mol,

$$|\Delta_r H^0| = \text{LHV} = \frac{802.31}{16.034} = 50.038 \text{ kJ/g} \approx 50 \text{ MJ/kg}$$

the value normally accepted.

A.4   The Entropy Balance

The entropy S is a property of any system, like the energy, and plays an important role in the calculations of the available energy (see Appendix B). Entropy can be transferred between systems by means of particular interactions (by heat transfer interactions) or by mass flow, and it is also generated by irreversibilities inside the systems. For a review of the numerous aspects and implications and interpretations of the concept of entropy, see [9].

In any case, taking the usual Fig. A.1 as a reference and assuming that:

1. The mean fluid velocities v_1 and v_2 on cross sections 1 and 2 are normal to A_1 and A_2
2. All the physical properties are uniform on the inlet and on the outlet sections 1 and 2
3. The viscous forces on section 1 and 2 are negligible
4. The entropy transported by conduction through A_1 and A_2 are negligible in comparison to the corresponding convective term
5. That it is possible to define locally all the required thermodynamic equilibrium properties of the fluid (P, T, U, S, etc.)

The entropy balance can be written as

$$\frac{d}{dt}(S_{tot}) = (\rho_1 A_1 v_1) S_1 - (\rho_2 A_2 v_2) S_2$$
$$+ \dot{S} + \dot{S}_G \tag{A.6}$$

in which

$$S_{tot} \equiv \int_{\mathcal{V}} \rho S d\mathcal{V} \qquad \text{the total entropy of the system}$$

$\dfrac{d}{dt}(S_{tot}) \equiv$ the time variation of the total entropy of the system

$(\rho_1 A_1 v_1) S_1 \equiv$ the entropy (convective) flow across cross section 1

$(\rho_2 A_2 v_2) S_2 \equiv$ the entropy (convective) flow across cross section 2

$\dot{S} \equiv \displaystyle\int_{\mathcal{S}} (\mathbf{n} \cdot \mathbf{s}) d\mathcal{S}$ rate of entropy into the fluid contained in the system

due to (molecular) heat transfer. The entropy flux $\mathbf{s} = \dfrac{\mathbf{q}}{T}$ is

equal to the local thermal energy flux divided by the local

temperature

$\dot{S}_G \equiv \displaystyle\int_{\mathcal{V}} \dot{s}_G d\mathcal{V}$ is the entropy production (the generated entropy)

into the system volume

The term $\dot{S}_G$ is the result of the contribution of different physical processes: (1) the entropy rate production associated with the internal heat transmission and (2) the entropy rate production from chemical reactions. The entropy rate $\dot{S}$, meanwhile, is the consequence of the entropy flux vector $\mathbf{s} = \mathbf{q}/T$ through the solid surfaces bounding the system. If the thermal power is exchanged with various sources/sinks of heat at different temperatures T_i, each constant, the integral relative to the term $\dot{S}$, may be replaced by a summation $\sum_i \dot{Q}/T_i$.

The balance of entropy represents a useful instrument for evaluating the work lost and the drop in conversion efficiencies due to irreversibilities (the term $\dot{S}_G$) within the fluid (the system) being considered. On this point, see Appendix B and C.

References

1. Macchi E (1979) Notes on thermo-fluid dynamics applied to machines. clup - cooperativa universitaria del politecnico, Milan (in Italian)
2. Allen III MB, Herrera I, Pinder GF (1988) Numerical modeling in science and engineering. Wiley, New York

3. Borel L (1984) Thermodynamique et Énergétique. Press Polytechniques Romandes, Lausanne, CH
4. Bird RB, Steward WE, Lightfoot EN (2002) Transport phenomena, 2nd edn. Wiley , New York
5. Gyftopoulos EP, Beretta GP (2005) Thermodynamics: foundations and applications. Dover Publications, Mineola, New York
6. Hangos KM, Cameron IT (2001) Process modelling and model analysis. Process systems engineering, vol 4. Academic Press, London, UK
7. Tester JW, Modell M (1977) Thermodynamics and its applications, 3rd edn. Prentice Hall PTR, Upper Saddle River, New Jersey
8. Mallard WG, Linstrom PJ (eds) (1998) NIST chemistry web book. NIST Standard Reference Database Number 69. National Institute of Standards and Technology, Gaithersburg MD http://webbook.nist.gov CitedDecember21st2011
9. Beretta GP, Ghoniem AF, Hatsopoulos GN (eds) (2008) Meeting the entropy challenge. An international thermodynamics symposium in honor and memory of Professor Joseph H. Keenan. MIT, 4–5 October 2007. American Institute of Physics, AIP Conferenze Proceeding 1033. Melville, New York

Appendix B
The Exergy or the Available Energy Function

The available energy (or work), understood in its widest sense, is a concept that has been discovered and rediscovered by many physicists and engineers over more than a century [1].

This concept of available energy helps to reveal the physical limit to the maximum useful energy that we can derive from the natural resources at our disposal, resources which are, at least in part, finite and, therefore, in our interest to conserve as much as possible.

All industrial processes take place in the earth's environment (atmospheric or marine), which can be considered an inexhaustible reservoir of energy, entropy, work, and matter. This is not really correct (for example, the earth's mass and energy are finite), but we are assuming that the global reservoir is enormous compared to any man-made system. Furthermore, we assume that energy (of any kind), work, and matter are released into the terrestrial environment at zero cost (in the case of pollution, which happens for the very reason that the global reservoir is not infinite, we can assume that the cleaning up operation is included in the industrial process, in such a way as to maintain the environment qualitatively unaltered).

Industrial processes exist as long as there are natural substances that are not in equilibrium with the earthly environment. These natural resources, in contrast with the environment, are attributed an economic value. In such a situation, though, the best we can do consists in bringing the thermodynamic state of a resource into balance with the terrestrial environment. Once this has happened, we cannot obtain anything further from our resource, since, with respect to the environment, it has reached a passive or dead state, that is, a mutually stable state of equilibrium. In any case, the concept of changing the thermodynamic state generally includes changing the nature of various substances via physical transformation or chemical and nuclear reaction.

The type of interaction between the system and the terrestrial environment depends on the walls placed between them: adiabatic or diathermic, rigid or movable, impervious or semipermeable or permeable. The work obtained or requested in the specific circumstances will depend on:

C.M. Invernizzi, *Closed Power Cycles*, Lecture Notes in Energy 11,
DOI 10.1007/978-1-4471-5140-1, © Springer-Verlag London 2013

- The thermodynamic characteristics of the terrestrial environment
- The system in consideration
- The kind of interaction allowed between the system and the environment
- The initial and final state of the system (the final state not necessarily coinciding with that of a mutually stable equilibrium with the environment)
- The irreversibilities of the process, which manifest themselves with a production of entropy

The terrestrial environment may reasonably be considered to be a reservoir (in a thermodynamic sense, see [2, p. 87], of work, heat, and mass at the constant temperature T_0, at the constant pressure P_0 and with the chemical potentials $\mu_{00,i}$ all constant. In effect, as we have said, the earth's environment is characterised by the fact of having huge values of internal energy, mass (that is, the number of moles $N_{00,i}$ of the various compounds that constitute it), and volume. However, it is a reservoir only in an approximate sense, given, for example, the fluctuations in temperature and pressure that affect it (see Sect. C.2 of Appendix C). Traditionally, it is transformed into a real reservoir by giving it the unalterable temperature of 25 °C and the pressure of a metric atmosphere (0.98 bar). The composition of its mass is still the subject of debate. Sciubba and Wall [1] discuss its specific aspects and lists a series of bibliographical references.

Once the reference environment has been defined, the exergy function (known also as the available energy function or availability function) can be defined as [1]:

> the maximum theoretical useful work obtained if a system is brought into thermodynamic equilibrium with the environment by means of processes in which the system interacts only with this environment.

With reference to Fig. B.1, in the absence of chemical reactions, we can then calculate the exergy associated with a fluid flow $\dot{m}$, available at pressure P_1 and temperature T_1 (on cross section 1) imagining a series of transformations that bring the fluid into a mechanical and thermal equilibrium with the environment (at temperature T_0 and pressure P_0), that is, into a restricted dead state, [3, p. 50], with W_{max} as output. We then imagine the separation of individual components, with the appropriate mass flows, by means of semipermeable membranes, and, finally, bring them into equilibrium with the environment from the point of view of their composition, too, by reaching a final state of complete balance (dead state) producing further useful power $\dot{W}^\star_{max} = \sum_i \dot{W}_i$. The exergy $\dot{E}$ will be the sum of the two terms: $\dot{E} = \dot{W}_{max} + \dot{W}^\star_{max}$.

The energy balance in stationary conditions (A.5), ignoring the changes in potential and kinetic energy, applied between the area sections 1 and 2 of Fig. B.1, gives

$$\dot{m}\,(H_1 - H_2) + \dot{Q} + \dot{W} = 0$$

The power $\dot{W}$ in the previous equation, derived from the formulation of (A.5), is, in this case, to be considered negative:

$$\dot{m}\,(H_1 - H_2) + \dot{Q} - \dot{W} = 0$$

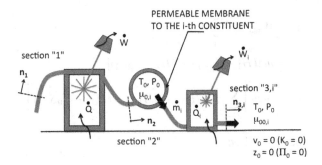

Fig. B.1 A schematic representation of a fluid system for calculating the exergy of a fluid with a defined composition, starting from the thermodynamic conditions of the cross section 1

The entropy balance (A.6) in stationary conditions gives

$$\dot{m}(S_1 - S_2) + \dot{S} + \dot{S}_G = 0$$

with S_1 and S_2 as specific entropies associated with the mass flow over sections 1 and 2, $\dot{S}$ as the entropy received from the fluid due to the heat exchange with the environment (at temperature T_0) and $\dot{S}_G$ as the entropy generated in the fluid through irreversibility.

Assuming $\dot{S} = \dot{Q}/T_0$, the combination of the energy and entropy balance gives the mechanical power $\dot{W}$:

$$\dot{W} = \dot{m}\left[(H_1 - H_2) - T_0(S_1 - S_2)\right] - T_0\dot{S}_G$$

with $H_1 = H(T_1, P_1, x_1, x_2 \cdots x_r)$, $S_1 = S(T_1, P_1, x_1, x_2 \cdots x_r)$, and $H_2 = H(T_0, P_0, x_1, x_2 \cdots x_r)$, $S_2 = S(T_0, P_0, x_1, x_2 \cdots x_r)$. Over cross section 1, the fluid is at room temperature and pressure, with the composition the same as at the start. The useful power $\dot{W}$ is at its maximum when $\dot{S}_G$ is null, that is,

$$\dot{W}_{max} = \dot{m}\left[(H_1 - T_0 S_1) - (H_2 - T_0 S_2)\right]$$

If we now imagine separating the i-th species by means of appropriate semipermeable membranes and bringing it up to the thermodynamic conditions $(T_0, P_{00,i}, \mu_{00,i})$, corresponding to those at which it is found in the environment, this gives a further power $\dot{W}^{\star}_{max,i}$ which, proceeding in the same way as before, is

$$\dot{W}^{\star}_{max,i} = \dot{m}x_i\left[(H_{2,i} - T_0 S_{2,i}) - (H_{3,i} - T_0 S_{3,i})\right]$$

with $H_{2,i} = H(T_0, P_0)$, $S_{2,i} = S(T_0, P_0)$ and $H_{3,i} = H(T_0, P_{00,i})$, $S_{3,i} = S(T_0, P_{00,i})$. The enthalpy $H_{2,i}$ and the entropy $S_{2,i}$ are to be considered as the enthalpy and the entropy of the i-th component when the fluid of which it is part (mix of gas, solution, etc.) is at a pressure P_0 and a temperature T_0. So, taking into account the presence of all the components, the exergy is

$$\dot{E} = \dot{W}_{max} + \sum_{i=1}^{r} \dot{W}_{max,i}^*$$

$$= \dot{m}\left[(H_1 - T_0 S_1) - (H_2 - T_0 S_2)\right]$$

$$+ \dot{m}\left[\sum_{i=1}^{r} x_i\,(H_{2,i} - T_0 S_{2,i}) - (H_{3,i} - T_0 S_{3,i})\right]$$

which may be generalised in

$$\dot{E} = \dot{m}\left[(H - T_0 S) - (H_0 - T_0 S_0)\right]$$

$$+ \dot{m}\left[\sum_{i=1}^{r} x_i\,(H_{0,i} - T_0 S_{0,i}) - (H_{00,i} - T_0 S_{00,i})\right] \tag{B.1}$$

In general, the term $[(H - T_0 S) - (H_0 - T_0 S_0)]$ is identified as "physical exergy" and the term $\left[\sum_{i=1}^{r} x_i\,(H_{0,i} - T_0 S_{0,i}) - (H_{00,i} - T_0 S_{00,i})\right]$ as "chemical exergy". For the elements, the term chemical exergy coincides with the exergy of the mixture of the i-th substance with the environment. In the case of chemical compounds, the chemical exergy must also take into account their formation exergy. In other words, the chemical exergy represents the work necessary to produce, in a reversible way, one unit of mass (kilogramme or mole) of the desired compound, taking its constituents from the environment.

If the enthalpies and the entropies refer to the mole, so too does the exergy $\dot{E}$ and the x_i ratios must be substituted by the respective molar fractions and the mass flow $\dot{m}$ represents the molar mass flow. In (B.1), the enthalpy H and the entropy S refer to the general thermodynamic state of the fluid we consider; T_0 is the temperature of the reference environment and the summation $\sum_{i=1}^{r} x_i\,(H_{00,i} - T_0 S_{00,i})$ is constant, once the reference environment and the fluid composition are set. Reference values for the chemical exergy of numerous elements and compounds are presented in [4–6].

B.1   The Exergy Balance

Still with reference to Fig. A.1 and assuming the same hypotheses used in writing the global balances for energy and entropy, (A.4) and (A.6) can be rewritten, introducing the specific enthalpy $H = U + P/\rho$ and extending them to the case of a system with several flows at input ($j = 1, 2 \cdots s$) and at output ($k = s + 1 \cdots t$). Ignoring the kinetic energies K and the potentials Π, we get

$$\frac{d}{dt}U_{\text{tot}} = \sum_{j(\text{in})=1}^{s} \dot{m}_j H_j - \sum_{k(\text{out})=s+1}^{t} \dot{m}_k H_k + \dot{Q} + \dot{W} \tag{B.2}$$

$$\frac{d}{dt}(S_{\text{tot}}) = \sum_{j(\text{in})=1}^{s} \dot{m}_j S_j - \sum_{k(\text{out})=s+1}^{t} \dot{m}_k S_k + \dot{S} + \dot{S}_G \tag{B.3}$$

The thermal power $\dot{Q}$ (according to the assumed conventions, or positive if it enter in the considered system) is defined by

$$\dot{Q} = \int_{\mathscr{S}} (\mathbf{n} \cdot \mathbf{q}) \, d\mathscr{S}$$

and the entropy $\dot{S}$, associated with the heat exchange on the surface $\mathscr{S}$ of the volume containing the fluid, evaluated by means of

$$\dot{S} = \int_{\mathscr{S}} (\mathbf{n} \cdot \mathbf{s}) \, d\mathscr{S}$$

$$= \int_{\mathscr{S}} \left(\mathbf{n} \cdot \frac{\mathbf{q}}{T} \right) d\mathscr{S}$$

can be related to one another by introducing the exergy $\dot{E}$ associated with the heat exchange. That is,

$$\dot{E}_Q = \int_{\mathscr{S}} (\mathbf{n} \cdot \mathbf{q}) \left(1 - \frac{T_0}{T} \right) d\mathscr{S}$$

$$= \int_{\mathscr{S}} (\mathbf{n} \cdot \mathbf{q}) \, d\mathscr{S} - T_0 \int_{\mathscr{S}} \left(\mathbf{n} \cdot \frac{\mathbf{q}}{T} \right) d\mathscr{S}$$

$$= \dot{Q} - T_0 \dot{S}$$

If the heat exchanges take place with heat sources or heat sinks at a constant temperature, the integrals extended to the surface $\mathscr{S}$ can be replaced by the appropriate summations and $\dot{E}_Q$ may be calculated as

$$\dot{E}_Q = \sum_l \dot{Q}_l - T_0 \sum_l \dot{S}_l$$

Substituting the previous ratio in the energy balance (B.2), we obtain

$$\frac{d}{dt}(U_{\text{tot}}) = \sum_{j(\text{in})=1}^{s} \dot{m}_j H_j - \sum_{k(\text{out})=s+1}^{t} \dot{m}_k H_k$$

$$+ \dot{E}_Q + T_0 \dot{S} + \dot{W}$$

Deriving $\dot{S}$ from (B.3) and substituting it in the previous relation give us

$$\frac{d}{dt} (U_{tot}) = \sum_{j(in)=1}^{s} \dot{m}_j H_j - \sum_{k(out)=s+1}^{t} \dot{m}_k H_k$$

$$+ \dot{E}_Q + T_0 \left(\frac{d}{dt} (S_{tot}) - \sum_{j(in)=1}^{s} \dot{m}_j S_j + \sum_{k(out)=s+1}^{t} \dot{m}_k S_k - \dot{S}_G \right) + \dot{W}$$

Or

$$\frac{d}{dt} (U_{tot} - T_0 S_{tot}) = \sum_{j(in)=1}^{s} \dot{m}_j \left(H_j - T_0 S_j \right) - \sum_{k(out)=s+1}^{t} \dot{m}_k \left(H_k - T_0 S_k \right)$$

$$+ \dot{W} + \dot{E}_Q - T_0 \dot{S}_G \tag{B.4}$$

The relation (B.4) makes it possible to calculate (setting $T_0 \dot{S}_G = 0$) the maximum useful work (if negative) or the minimum work necessary (if positive) $\dot{W}_{optimal}$ when the flows (or flow) under consideration, interacting with one another and with the machinery present in volume $\mathcal{V}$ which delimits the system, change from thermodynamic conditions that are "in" to those which are "out". It only takes into account the variations in physical exergy. Because of the usual conventions, the mechanical power $\dot{W}$ will be considered as supplied to the system.

The exergetic flows could also take into account the quota of chemical exergy, namely, the work corresponding to the attainment of the dead state by all the flows at both outlet and inlet. Or assuming

$$\dot{E}_j = \dot{m}_j \left(H_j - T_0 S_j \right)$$

$$= \dot{m}_j \left[\left(H_j - T_0 S_j \right) - \left(H_{0,j} - T_0 S_{0,j} \right) \right]$$

$$+ \dot{m}_j \sum_{i=1}^{r} x_{ji} \left[\left(H_{0,ji} - T_0 S_{0,ji} \right) - \left(H_{00,ji} - T_0 S_{00,ji} \right) \right]$$

$$\dot{E}_k = \dot{m}_k \left(H_k - T_0 S_k \right)$$

$$= \dot{m}_k \left[\left(H_k - T_0 S_k \right) - \left(H_{0,k} - T_0 S_{0,k} \right) \right]$$

$$+ \dot{m}_k \sum_{i=1}^{r} x_{ki} \left[\left(H_{0,ki} - T_0 S_{0,ki} \right) - \left(H_{00,ki} - T_0 S_{00,ki} \right) \right]$$

In which case, $\dot{W}_{optimal}$ is then calculated by assuming $\dot{E}_Q = 0$. In short, referring to stationary conditions,

$$\dot{W}_{\text{optimal}} = -\sum_{j(\text{in})=1}^{s} \dot{E}_j + \sum_{k(\text{out})=s+1}^{t} \dot{E}_k \tag{B.5}$$

The term $T_0 \dot{S}_G$ in (B.4) represents the exergy lost in the unit time through irreversibilities present in the system in question. The exergy, like the entropy, is not preserved, and undesired, but inevitable, increases in the entropy cause reductions (losses) of exergy with consequent reduction or increase in the power compared to the ideal value $\dot{W}_{\text{optimal}}$. In Appendix C, the exergetic balance is used in order to analyze the irreversibilities in thermodynamic cycles and in some thermal machinery.

Exercises

B.1. We calculate the minimum power (the exergy rate) necessary to produce one kilogramme of oxygen, separating it from the dry air.

We consider the dry air as a mixture of ideal gases, consisting of oxygen (present with a molar fraction of 21 %) and nitrogen (molar fraction of 79 %). The exergy may be calculated by using (B.5) under stationary conditions, imagining that the process takes place at room temperature T_0, starting with a room pressure of P_0 and assuming $T_0 \dot{S}_G = 0$.

In this case, $j = 1$, $k = 2, 3$, and $i = 1, 2$. Making $\dot{m}_1 = \dot{m}_{\text{air}}$, $\dot{m}_{11} = \dot{m}_1 x_{11} = \dot{m}_{O_2}$ and $\dot{m}_{12} = \dot{m}_1 x_{12} = \dot{m}_{N_2}$. The final product is an oxygen mass flow $\dot{m}_{21} = \dot{m}_{O_2} = 1$ kg/s, with $\dot{m}_{32} = \dot{m}_{N_2}$. The molar fractions of the air at inlet are $y_{11} = y_{O_2} = 0.21$ and $y_{12} = y_{N_2} = 0.79$. The mass fractions of the air are $x_{11} = y_{11} M_{11}/M = 0.21(32)/28.84 = 0.233$ and $x_{12} = y_{12} M_{12}/M = 0.79(28)/28.84 = 0.767$. The air flow to be treated, assuming that all the oxygen contained in it has been separated, equals $\dot{m}_1 = \dot{m}_{21}/x_{11} = 1/x_{11} = 4.762$ kg/s.

The exergies E_{21} and E_{32} are

$$E_{21} = (H_{0,21} - T_0 S_{0,21}) - (H_{00,21} - T_0 S_{00,21})$$
$$= (H_{0,21} - H_{00,21}) - T_0 (S_{0,21} - S_{00,21})$$
$$= -T_0 R \ln y_{11}$$
$$= -T_0 (259.8) \ln 0.21 = +T_0 (405.46) \text{ J/kg}$$

$$E_{32} = (H_{0,31} - T_0 S_{0,31}) - (H_{00,31} - T_0 S_{00,31})$$
$$= (H_{0,31} - H_{00,31}) - T_0 (S_{0,31} - S_{00,31})$$
$$= -T_0 R \ln y_{32}$$
$$= -T_0 (296.9) \ln 0.79 = T_0 (69.99) \text{ J/kg}$$

The exergies E_{11} and E_{12} are null: the natural components of the air (in physical and chemical balance with the environment) cannot supply any useful work.

From the macroscopic exergy balance expressed by (B.5),

$$\dot{W}_{\text{optimal}} = (\dot{m}_{21}E_{21} + \dot{m}_{32}E_{32})$$
$$= T_0 (405.46 + (3.762) 69.99)$$
$$= T_0 (668.76) \text{ W/kg/s of oxygen.}$$

With $T_0 = 298.15$ K, we get $\dot{W}_{\text{optimal}} = 199$ kW per kg/s of produced oxygen.

B.2. We calculate the power ideally available when a mass flow $\dot{m}_1 = \dot{m}_{H_2O} = 1$ kg/s of pure water flows into the sea.

We suppose that the sea water consists of just pure water and (NaCl) with a concentration of salt equal to 35 g/kg [7, p. 31]. The temperature at which the mixing occurs is T_0 (298.15 K) and the pressure is that of the environment (P_0).

In this particular case, $\dot{W}_{\text{optimal}}$ equals the chemical exergy rate of the freshwater, calculated when considering the sea as the reference environment. Hypothesising that the sea water is an ideal solution with the molar composition $y_1 = y_{H_2O} = (0.965/18.015)/(0.965/18.015+0.035/58.5) = 0.989$ and $y_2 = y_{NaCl} = 1-y_1 = 0.011$, from (B.5),

$$\dot{W}_{\text{optimal}} = - (-\dot{m}_1 T_0 R \ln y_1) = T_0 R \ln 0.989 = -T_0 R (0.011) \text{ W}$$

Or $\dot{W}_{\text{optimal}} = -1.52$ kW per kg/s of pure water. The negative value obtained, in accordance with our conventions, indicates that the work is useful.

B.3. We consider an OTEC system[1] using hot water at temperature $T_1 > T_0$ (see Fig. B.2) and, contemporarily, exploiting the exergy available in the water, which is colder than its surrounding environment, at temperature $T_2 < T_0$. We calculate the maximum power extractable from a flow of hot water $\dot{m}_1 = 1$ kg/s.

In this case, we consider only the physical exergy as being useful. Referring to Fig. B.2, we imagine, in system A, cooling the hot water from temperature T_1 to temperature T_3. In this way, a useful power $\dot{W}_{\text{optimal},A}$ is generated, at the same time as a thermal power $\dot{Q}_{\text{min}}$ is released into the environment. This power, at temperature T_0, can produce (in system B) a further mechanical power $\dot{W}_{\text{optimal},B}$ by exploiting the availability of cold water at temperature T_2. We assume $(T_1 - T_3) = (T_4 - T_2)$.

---

[1]The practice of exploiting the natural thermal gradients that arise in the oceans is called OTEC (Ocean Temperature Energy Conversion). Where the water is deep, the highest temperature is to be found in the surface layers, whilst the deeper water stays colder. In tropical waters, the surface temperature may even reach temperatures that are $20 \div 25$ °C higher than the water at depth. Such a temperature difference, when constant throughout the year, may be used to produce mechanical energy by means of a heat engine. It was Jacques Arsene d'Arsonval (a French physicist) who, in 1881, first proposed the use of the ocean's temperature gradient as a means of producing electricity. One of his students, George Claude, built the first OTEC plant, in Cuba in 1930.

Fig. B.2 Conceptual scheme for calculating the exergy available in an OTEC system. The surrounding environment is at temperature T_0 and pressure P_0

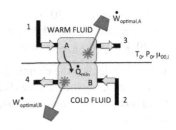

From the exergetic balance (B.4), applied to subsystem A, we get $\dot{W}_{max,A}$ (with $T_0\dot{S}_G = \dot{E}_Q = 0$ and in stationary conditions)

$$\dot{W}_{optimal,A} = -\dot{m}_1\left[(H_1 - H_3) - T_0(S_1 - S_3)\right]$$

Thermal power $\dot{Q}_{min}$ not converted into useful work and discharged into the environment at temperature T_0 is (from the balance of powers B.2 under stationary conditions)

$$\dot{Q}_{min} = -\dot{m}_1 T_0(S_1 - S_3)$$

The result is a flow $\dot{m}_2$, equal to

$$\dot{m}_2 = \dot{m}_1 T_0 \frac{S_1 - S_3}{H_4 - H_2}$$

From the exergetic balance (B.4) applied to subsystem B, we get $\dot{W}_{optimal,B}$ (with $T_0\dot{S}_G = \dot{E}_Q = 0$ and in stationary conditions)

$$\dot{W}_{optimal,B} = \dot{m}_2\left[(H_4 - H_2) - T_0(S_4 - S_3)\right]$$

And the total power available, in accordance with (B.5), is

$$\dot{W}_{optimal} = \dot{W}_{optimal,A} + \dot{W}_{optimal,B}$$
$$= -\dot{m}_1\left[(H_1 - H_3) - T_0(S_1 - S_3)\right]$$
$$+ \dot{m}_2\left[(H_4 - H_2) - T_0(S_4 - S_2)\right]$$

Or assuming constant specific heats for the two flows of hot and cold water,

$$\dot{W}_{optimal} = (-1)\dot{m}_1(H_1 - H_3)\left(1 - 2T_0\frac{S_1 - S_3}{H_1 - H_3} + T_0^2\frac{S_1 - S_3}{H_1 - H_3}\frac{S_4 - S_2}{H_4 - H_2}\right)$$
$$= (-1)\dot{m}_1 C_P(T_1 - T_3)\left(1 - 2T_0\frac{\ln T_1/T_3}{T_1 - T_3} + T_0^2\frac{\ln T_1/T_3}{T_1 - T_3}\frac{\ln T_4/T_2}{T_4 - T_2}\right)$$

Assuming $T_1 = 30\,°C$, $T_3 = 25\,°C$, $T_2 = 5\,°C$, and $T_4 = 10\,°C$. With $T_0 = 298.15\,K$ and $\dot{m}_1 = 1\,kg/s$, we get $\dot{W}_{optimal} = -C_P\,(5)\,0.07 = -(4.0)\,0.351 = -1.40\,kW$ per kg/s of warm sea water.

B.4. We calculate the chemical exergy of the methane.

We can apply the balance (B.5), considering the chemical exergy of the methane as an unknown and assuming $\dot{W}_{optimal}$ equal to the exergy in the formation reaction of the methane:

$$C\,(s) + 2H_2\,(g) \rightarrow CH_4\,(g)$$

The exergy $\dot{W}_{optimal}$ is derived from the balance (B.4) with $\dot{E}_Q = 0$ and $T_0\dot{S}_G = 0$, at pressure P_0 and temperature T_0 and introducing the enthalpies and entropies of formation.

In the specific case, [8]: $\Delta_f H^0_{C(s)} = 0$, $\Delta_f H^0_{O_2(g)} = 0$, $\Delta_f H^0_{CH_4(g)} = -74.87\,kJ/mol$, $S^0_{C(s)} = 5.833\,J/mol\,K$, $S^0_{H_2(g)} = 130.68\,J/mol\,K$, $S^0_{CH_4(g)} = 188.66\,J/mol\,K$.

Thus, with reference to a mole of methane per second, from the balance (B.4),

$$-T_0\,(0.005833) - 2T_0\,(0.13068) - [(-74.87) - T_0\,(0.18866)] + \dot{W}_{optimal} = 0$$

$$74.87 - T_0\,(0.078533) + \dot{W}_{optimal} = 0$$

Or $\dot{W}_{optimal} = -51.467\,kJ/mol$. Then, introducing the chemical exergies (the standard reference exergies) from the balance (B.5),

$$\dot{W}_{optimal} = -\dot{E}_{C(s)} - 2\dot{E}_{H_2(g)} + \dot{E}_{CH_4(g)}$$

$$\dot{W}_{optimal} + \dot{E}_{C(s)} + 2\dot{E}_{H_2(g)} = \dot{E}_{CH_4(h)}$$

The standard exergies of the compounds and elements have been calculated (along the lines of what was done in Exercise B.1) from various authors. In this case, for methane, assuming $\dot{E}_{C(s)} = 410.26\,kJ/mol$ (from [6, p. 398, Table A.20]) and $\dot{E}_{H_2(g)} = 236.1\,kJ/mol$ (from [6, p. 189, Table 6.1]),

$$\dot{E}_{CH_4(g)} = \dot{W}_{optimal} + \dot{E}_{C(s)} + 2\dot{E}_{H_2(g)}$$

$$= (-51.467) + 410.26 + 2\,(236.1)$$

$$= 830.99\,kJ/mol\ of\ CH_4.$$

B.5. Still with reference to methane, we evaluate the exergy of combustion. That is, the maximum work available from the combustion of methane.

We consider the stoichiometric combustion, with just oxygen present, at temperature T_0 and pressure P_0. The reference reaction is

$$CH_4\,(g) + 2O_2\,(g) \rightarrow CO_2\,(g) + 2H_2O\,(g)$$

The balance (B.5) (in stationary conditions), applied to the chemical reaction of the combustion, gives (assuming the exergy from the oxygen has the value from [6, p. 189, Table 6.1] and the chemical exergy from the water has the value from [6, p. 384, Table A.13])

$$\dot{W}_{optimal} = -\dot{E}_{CH_4(g)} - 2\dot{E}_{O_2(g)} + \dot{E}_{CO_2(g)} + 2\dot{E}_{H_2O(g)}$$
$$= -830.99 - 2\,(3.97) + 19.9 + 2\,(9.5)$$
$$= -800.03 \text{ kJ/mol of } CH_4.$$

an absolute value of little less than the lower heating value of -802.31 kJ/mol (see Exercise A.9).

B.6. The complete combustion reaction of methane in air

$$CH_4\,(g) + 2\,(O_2\,(g) + 3.762N_2\,(g)) \rightarrow CO_2\,(g) + 2H_2O\,(g) + 7.524N_2\,(g)$$

requires 2 moles of oxygen for every mole of methane. That is, for the complete oxidation of one unit mass of methane (the fuel), 17.108 kg of air are needed: the stoichiometric air-fuel mass ratio (AFR).[2]

In correspondence with the stoichiometric AFR we obtain the maximum adiabatic flame temperature,[3] which decreases with respect to the maximum value, both in conditions of excessive air (when there is a large quantity of nitrogen and oxygen in excess in the combustion products) and in shortages of air (when the oxygen present is insufficient for oxidising all the fuel) (see Fig. B.3a).

The ratio between the physical exergy difference (that is, the maximum ideal work obtainable when the combustion products have cooled from the adiabatic flame temperature to room temperature T_0) and the LHV of the methane is in Fig. B.3b. The maximum value of the ratio (about 0.72) is reached under stoichiometric conditions. Since the lower heat value (see Problem B.5), almost always coincides (to within a few percentage points) with the exergy of the fuel, the combustion is responsible for a minimum loss of about 30 % of the fuel's initial energy availability. The exergetic loss increases with the air-fuel ratio. However, technical reasons often make it necessary to operate with an excess of air, sometimes a substantial excess. In gas-turbine engines, e.g., it is common to have AFR

---

[2]For the calculations in this example, the dry air has been assimilated into an ideal mixture consisting of oxygen (with a molar fraction of 21 %) and nitrogen (with a molar fraction of 79 %). In this hypothesis, each oxygen molecule accompanies 3.762 nitrogen molecules.

[3]The adiabatic flame temperature has been calculated assuming chemical equilibrium, with the adiabatic combustion chamber at constant pressure P_0 and not taking into account the formation mechanisms of the CO and NO. In reality, the carbon monoxide CO forms naturally for AFRs below the stoichiometric level and the nitrogen oxide NO forms through thermal dissociation (at high temperature) of the nitrogen present in the air or from the nonmolecular nitrogen that may be present in the fuel.

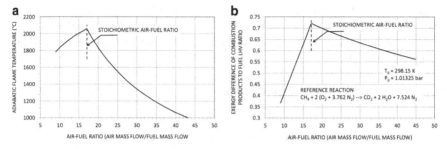

Fig. B.3 (**a**) Adiabatic flame temperature for the combustion reaction of methane in air, with the variation in the air-fuel ratio. Calculation at the chemical equilibrium. Combustion completed. (**b**) Ratio between the variations in exergy of the adiabatic combustion products from methane in air, with varying air-fuel ratio. The combustion products, available at pressure P_0 and adiabatic flame temperature, are cooled down to temperature T_0

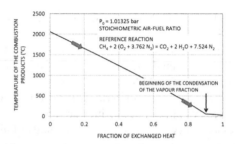

Fig. B.4 Temperature of methane combustion products as they cool down to temperature T_0

values that are 2–3 times the stoichiometric value, in order to prevent excessive temperatures at the end of combustion, keeping them below 1,200–1,300 °C.

Figure B.4 represents the temperature trend of combustion products during their cooling down to room temperature T_0. At a temperature of around 60 °C, the fraction of steam contained in the gases begins to condense, making its own condensation heat available.

In theory, further work is available as chemical exergy of the compounds of the combustion products, once they are available at the pressure P_0 and temperature T_0.

The chemical exergy is calculated by imagining the separation of the chemical species present in the combustion products (available at pressure P_0 and temperature T_0) and mixing them then with the atmosphere. We also need to take into account the fraction of condensed steam. However, since the contribution of the chemical exergy is usually fairly modest compared to the quota of physical exergy available (and, in fact, is unusable), it is not considered.

B.7. We calculate the minimum work needed to separate the carbon dioxide from the combustion products of coal.

The techniques for trapping the carbon dioxide originating from burning fossil fuels (typically coal, but also methane) today constitute a rapidly growing field of research and development. By "sequestration of CO_2", we mean capturing the carbon dioxide in the power station in such a way as to prevent it being released into the atmosphere, transporting it through pipelines and its ultimate geological storage. Generally, there are many polluting compounds deriving from coal combustion: nitrogen oxides, sulphur oxides, powders, and even very low concentrations (a few parts per billion), of highly toxic metals. Fluorine and chlorine are also frequently present. We presume that in the final products of the combustion, these pollutants have been completely removed, and as reference, we consider the combustion of coal (graphite) in air with a stoichiometric AFR:

$$C(s) + (O_2(g) + 3.762N_2(g)) = CO_2(g) + 3.762N_2(g)$$

After the separation, we assume that the CO_2 must be available in liquid phase, at temperature T_0 of 25 °C and a pressure of 150 bar. After all the necessary treatments and the preheating of the combustion air, the combustion products (CO_2 and nitrogen) are available at the chimney at the pressure P_0 (1 bar) and the temperature of 130 °C.

We break the process down into three phases: (A) cooling of the mixture of carbon dioxide and nitrogen, at room pressure, from a temperature of 130 °C down to a temperature T_0 of 25 °C. For this first phase, we calculate the optimal work $\dot{W}_{optimal,A}$ from (B.4); (B) separation of the CO_2 from the nitrogen, operating at a temperature T_0 and pressure P_0. From (B.5), we calculate $\dot{W}_{optimal,B}$; (C) the isothermal compression at temperature T_0 until the final pressure of 150 bar. From (B.4), we calculate the $\dot{W}_{optimal,C}$. The (algebraic) sum $\dot{W}_{optimal,A} + \dot{W}_{optimal,B} + \dot{W}_{optimal,C}$ will give the optimal global work.

For 1 kmol/s of coal (12.011 kg/s), we get a mass flow $\dot{m}$ of combustion products of 149.3966 kg/s, of which 44.0098 kg/s are CO_2 and 105.3868 kg/s di N_2. The molar composition of the gases produced by the combustion gives $y_{CO_2} = 0.21$ and $y_{N_2} = 0.79$.

For the phase (A),

$$\dot{W}_{optimal,A} = -\dot{m}(\Delta H - T_0 \Delta S) = -2262.2752 \text{ kW}$$

For the second phase (B), $j = 1, k = 2, 3$ e $i = 1, 2$ (two components). Con $\dot{m}_{11} = \dot{m}x_{11} = \dot{m}_{CO_2}$ and $\dot{m}_{12} = \dot{m}x_{12} = \dot{m}_{N_2}$. For the products $\dot{m}_{21} = \dot{m}_{CO_2}$ and $\dot{m}_{32} = \dot{m}_{N_2}$, and the balance (B.5) becomes

$$\dot{W}_{optimal,B} = -(\dot{m}_{11}E_{11} + \dot{m}_{12}E_{12}) + (\dot{m}_{21}E_{21} + \dot{m}_{32}E_{32})$$

with, in the hypothesis that the gas mixture can be considered a mixture of ideal gases, $E_{11} = (T_0 R \ln y_{11} + E_{CO_2(g)})$, $E_{12} = (T_0 R \ln y_{12} + E_{N_2(g)})$, $E_{21} = E_{CO_2(g)}$ and $E_{32} = E_{N_2(g)}$. So,

$$\dot{W}_{optimal,B} = -(\dot{m}_{11}T_0 R \ln y_{11} + \dot{m}_{12}T_0 R \ln y_{12}) = 6067.0075 \text{ kW}$$

For the final phase (C),

$$\dot{W}_{\text{optimal},C} = -\dot{m}_{21}\left(\Delta H - T_0 \Delta S\right) = 9819.0574 \text{ kW}$$

Therefore, the power necessary to separate the carbon dioxide completely is $\dot{W}_{\text{optimal}} = (-2262.2752 + 6067.0075 + 9819.0574)/44.0098 = 309.56 \text{ kW}/(\text{kg/s})$ of carbon dioxide available at the end of the process in liquid form at 25 °C and 150 bar.

The energy costs for the sequestration process are significant, and in the real case, the net efficiency of a thermoelectric power station using such a process would fall by several percentage points (5–8), with an additional power consumption of around 10–20 % of the net power produced and a sizeable increase in the cost of generating electricity (in certain cases, as much as 50 % more, [9, 10], depending on the type of plant, the type and cost of fossil fuel used and the technology employed in separating the carbon dioxide).

References

1. Sciubba E, Wall G (2007) A brief commented history of exergy from the beginnings to 2004. Int J Therm 10(1):1–26
2. Gyftopoulos EP, Beretta GP (2005) Thermodynamics: foundations and applications. Dover Publications, Mineola, New York
3. Galliani A, Pedrocchi E (2006) Exergy analysis. Polipress, Polytechnic of Milan, Milan. (in Italian)
4. Morris DR, Szargut J (1986) Standard chemical exergy of some elements and compounds on the planet Earth. Energy 11(8):733–755
5. Szargut J (1989) Chemical exergies of the elements. Appl Energ 32:269–285
6. Delgado AV (2008) Exergy evolution of the mineral capital on earth. Ph.D. Thesis. Department of Mechanical Engineering, Centro Politécnico Superior University of Zaragoza, Zaragoza
7. Cerci Y, Cengel Y, Wood B, Khraman N, Karakas ES (2003) Improving the thermodynamic and economic efficiencies of desalination plants: minimum work required for desalination and case studies of four working plants. Agreement No. 99-FC-81-0183. Mechanical Engineering University of Nevada. Reno, Nevada
8. Mallard WG, Linstrom PJ (eds) (1998) NIST chemistry web book. NIST Standard Reference Database Number 69. National Institute of Standards and Technology, Gaithersburg MD http://webbook.nist.govCitedDecember21st2011
9. Lozza G, Chiesa P, Romano M, Valenti G (2009) CO_2 capture from natural gas combined cycles. In: Proceedings of the Sustainable Fossil Fuels For Future Energy Conference (S4FE 2009), Roma, July 2009
10. Kreutz T, Williams R, Consonni S, Chiesa P (1989) Co-production of hydrogen, electricity and CO_2 from coal with commercially ready technology. Part B: Economic analysis. Int J Hydrogen Energ 30:766–784

Appendix C
Irreversibilities in the Thermodynamic Cycles and in the Thermal Machines[1]

C.1 Irreversibilities, Both Internal and Towards the External Environment

Preliminary analysis of the conversion cycles is normally carried out with simplified hypotheses, used to help define the ideal reference cycles, where the machines will operate without losses (reversibly) and any internal heat exchange takes place at the slightest change in temperature.

As an example, Fig. C.1 represents the Rankine, Brayton, Otto and Diesel cycles, characterised by the same external temperatures (maximum and minimum) and where efficiency (ratio between useful work and heat introduced at high temperature), for the working parameters shown in the figure, vary between 61 % in the Diesel cycle and 37 % in the Brayton cycle.

This widely differing energy efficiency is fully justified by the different modes of introducing the primary heat and the rejection of the discarded heat: in a strict sense, these cycles can only be considered "ideal" if they are related to an environment that supplies primary heat exactly according to the law by which the working fluid requests it and that makes available a cooling capacity in accordance with the effective law of the rejection of discarded heat. In other words, the varying energy efficiencies are merely due to a different external environment making the conversion cycle completely reversible. Consequently, prior to any detailed analysis, consideration must be given to the environment surrounding the prospective thermodynamic cycles.

---

[1]The following appendix is based on the text of a lesson prepared in March 1996 by Prof. Gianfranco Angelino (1938–2010) and the author of this book, Costante M. Invernizzi, but never published.

C.M. Invernizzi, *Closed Power Cycles*, Lecture Notes in Energy 11,
DOI 10.1007/978-1-4471-5140-1, © Springer-Verlag London 2013

Fig. C.1 Ideal steam and gascycles with the same extreme temperatures

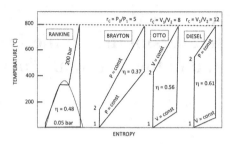

C.2 Definition of the Characteristics of the Cooling Well and the Heat Source

If we sought to make our analysis as realistic as possible, each time and for each cycle, we would have to define the relevant cooling environment and a high-temperature heat source, in order to evaluate not only the internal imperfections but also the irreversibility towards the outside world. Such a definition, though, would not only be difficult from a practical point of view, but also a conceptual one, which makes a little clarification appropriate.

The Cooling Well

The environment in which human activity takes place and which ought to constitute the "zero" state for all forms of energy, in reality, exhibits a wide variability, of various kinds, which makes it far more complex and articulated than shown in the usual thermodynamic simplifications. Here below, we give several examples of this disuniformity:

- Temperature differences between the various constituents. Air and water show marked differences in temperature, also in terms of behavior over time. Water in deep basins reveals a notable temperature variation, which can lend itself to practical uses. For example, in the OTEC systems [1–4], a temperature difference of 22–23 °C between the surface water in tropical regions and that at great depth (for example, below 700 m) is used in part to sustain a thermodynamic cycle and, in part, to provide, as sensible heat, an adequate heat flow at both input and output. In Italy, the sea surface temperatures have a seasonal variability of 10–12 °C (for example, a winter minimum of 13 °C and a summer maximum of 25 °C). At depths greater than 100 m, the temperature remains more or less constant at the winter surface temperatures [5, 6]. The use of deep water for cooling in thermoelectric power stations, where technically feasible, would mean an additional difference of ten degrees available in the summer period.
- Lack of equilibrium between the humidity content of air under effective conditions and in saturation. By way of example, without considering extreme situations of disuniformity, like desert climates, in the Milanese hinterland, the

average relative humidity in the month of July is 59 %, with a corresponding average temperature of 24 °C, [7]. This corresponds to a saturation temperature (theoretically available by the refrigeration effect of vaporising the water) of 15.3 °C and to a wet bulb temperature of 18.5 °C. The disuniformity in steam concentration, transformed into thermal motive power, leads so to temperature differences of around ten degrees for summers in the Lombard plain.

- Cyclical temperature variations on a yearly or daily basis. Taking the Milan area as an example again, compared to an average annual value of about 13 °C (measured constantly deep in the ground and, less reliably, in the aquifer waters), there are annual maximum and minimum values for the atmospheric temperature of 30 °C (average of the maximums for July) and of −1.5 °C (average of the minimums for January).

One example of the benefit to be derived from this temperature excursion is provided by an experimental prototype for a steam cycle, where the low-pressure section is replaced by an ammonia recovery cycle, which dissipates the discharged heat into the atmosphere by means of an air cooler [8]. The high saturation pressure of the ammonia, even at low temperature, allows the complete refrigeration capacity of the atmosphere to be used, even at the lowest seasonal temperatures. The potential importance of daily fluctuations becomes clear when considering that the summer temperature excursion between day and night (10–12 °C) is such as to consent the (statistical) solution of any problem of summer cooling, by resorting to simple systems of accumulation.

- Differences in saline concentrations between flowing freshwater and sea water. The work theoretically available from the reversible mixing of 1 m^3 of freshwater with sea water with a concentration of 3.5 % is around 0.4 kWh, which is about three times more than the work available in an ideal OTEC system as described, from 1 m^3 of warm surface water.[2]

These disuniformities give the designer the opportunity to choose the most favorable technical situation offered by the "imperfections" of the "zero" state. Furthermore, the fluctuations may not be great in absolute terms, but they have a fundamental importance for energy systems using low density energy (heat recovery, geothermal energy, solar heat, etc.). Given the variability of the environment, the choice of a "zero" state will be conventional in any case. Consequently, the effective environment is usually replaced by an isotherm at the minimum temperature of the cycle in question.

The Heating Reservoir

The effective energy source for thermodynamic cycles usually consists of a flow of fuel and air, and the transfer of heat to the working fluid requires a combustion

---

[2]By way of comparison, the work obtained from 1 m^3 of water after a fall of 100 m is about 0.3 kWh.

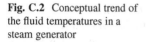

Fig. C.2 Conceptual trend of
the fluid temperatures in a
steam generator

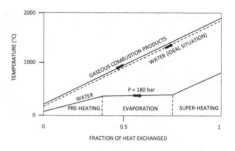

process before the fuel energy is available as sensible heat in the combustion by-products. Consequently, there are two irreversibilities, one after the other: that of the combustion and that of the heat transfer under finite temperature differences. The first of these has been analyzed in Sect. B.1, Problem B.5, whilst the meaning and the implications of the second ought to be considered, before examining the details.

To this end, the diagram in Fig. C.2 represents the concept of the heat exchange process in a steam generator (the actual organisation of the heat exchange would be fairly complex). The enormous irreversibility of the heat exchange is the consequence of the great temperature differences between the hot gases and the water. This could be nullified, without modifying the combustion process, by altering the water cycle at depth, in such a way that it receives the heat from the gas at the temperature at which it is available (dotted line). In this way, provided it is internally reversible, the cycle would produce the maximum work. Attainment of this extreme useful effect would be the consequence of two different operations: the thermodynamic refinement of the cycle and the solution (unrealistic today) of the metallurgical problems connected with the creation of a high-pressure heat exchanger for very high temperatures. Such an optimal efficiency could have an ambiguous significance, in that it would not be clear which part is derived thermodynamically (with current limitations on materials) and which part is due to metallurgical progress.

In order to distinguish between the two problems, the effective heat source could be replaced by an isothermal source, at the maximum temperature of the cycle. In that way, any deviation in the conversion efficiency from a Carnot cycle would be due exclusively to a thermodynamic insufficiency in the cycle. Despite being rather conventional, this choice would separate the thermodynamic problems from those of combustion, which, all things considered, only marginally affects the quality of the conversion processes.

C.3   Introduction to the Thermodynamic Analysis of Losses

Once a heat source and a reference environment have been defined, as suggested in the previous section, all the reversible conversion cycles are equally perfect, since they produce the maximum useful effect. Any reversibility generates a loss in the final balance of the work, which can be quantified by a relationship (see (B.4), in stationary conditions):

$$\Delta W_i = \Delta E_i = T_0 S_{G,i} \tag{C.1}$$

where the index i represents the generic location of the irreversibility, T_0 is the temperature of the environment and $S_{G,i}$ is the production of entropy caused by the irreversibility. Equation (C.1) rewritten in terms of efficiency loss becomes (see (1.22))

$$\Delta \eta_i = \frac{T_0}{Q_{in}} S_{G,i} \tag{C.2}$$

in which Q_{in} is the primary heat introduced into the cycle. If the significant components m that create the conversion cycle are each characterised by a parameter of merit ϵ_j (for example, compression or expansion efficiency, regeneration efficiency), the final efficiency expression is (see (1.22))

$$\eta = \eta_{max} - \sum_{i=1}^{n} \Delta \eta_i \tag{C.3}$$

with

$$\Delta \eta_i = \Delta \eta_i \left(\epsilon_1, \epsilon_2, \cdots \epsilon_m \right) \tag{C.4}$$

Equation (C.4) should be interpreted in the sense that the loss occurring in the i-th process is the outcome of the way in which all the components operate. If this corresponds to what happens in reality, then the loss $\Delta \eta_i$ or the entropic production $S_{G,i}$ which concretises it does not have an exclusive causal relationship with the quality of the i-th component. The entropic analysis may reveal the site, even by accident, but not the cause of a loss. This can be made clearer with an example. Consider the dissipation of kinetic energy K at the outlet of a gas turbine at temperature T. The corresponding work loss equals $\Delta W'_K = T_0 K / T$ and represents a fraction of K. The internal energy of the discharge gases increases by the quantity K which, once released definitively, in the form of heat to the environment, gives rise to a second loss $\Delta W''_K = T_0 (K / T_0 - K / T)$. The total loss $\Delta W'_K + \Delta W''_K$ is equal to K. The value $\Delta W'_K$ underestimates the real effect of the dissipation of kinetic energy K because an important secondary effect $(\Delta W''_K)$ is transferred to another site but is causally connected with the first phenomenon. Generally speaking, then, it is necessary to consider not only the direct effects of the losses but also the indirect ones.

This receives further clarification in the case where the entropic loss calculated by (C.1) means the minimum loss, unrecoverable and connected to a given irreversible process, as would appear, without additions, if the irreversibility in question did not induce others within the overall thermodynamic process.

The entropic analysis of the losses conforms to the definition, for each transformation, of a second-law efficiency that, for the components currently in use, assumes the following expression:

Fig. C.3 Adiabatic efficiency and second-law efficiency as a function of the polytropic efficiency of compression

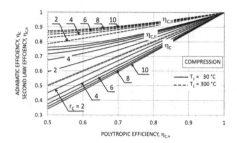

Fig. C.4 Adiabatic efficiency and second-law efficiency as a function of the polytropic efficiency of expansion

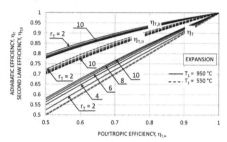

Adiabatic compressor

$$\eta_{C,II} = \frac{W - \Delta E}{W} = \frac{W - T_0 S_G}{W} \tag{C.5}$$

which can also be used for an adiabatic pump;

Adiabatic turbine

$$\eta_{T,II} = \frac{W}{W + \Delta E} = \frac{W}{W + T_0 S_G} \tag{C.6}$$

while for a *heat exchanger* in which a hot fluid is cooled from $T_{I,\text{in}}$ to $T_{I,\text{out}}$ and a cold fluid heated from $T_{II,\text{in}}$ to $T_{II,\text{out}}$

$$\eta_{HE,II} = \frac{1 - \dfrac{T_0}{T_{II,\text{out}}-T_{II,\text{in}}} \ln \dfrac{T_{II,\text{out}}}{T_{II,\text{in}}}}{1 - \dfrac{T_0}{T_{I,\text{in}}-T_{I,\text{out}}} \ln \dfrac{T_{I,\text{in}}}{T_{I,\text{out}}}} \tag{C.7}$$

obtained with the hypothesis of constant specific heat for both fluids.

The existing correlation between second-law efficiency η_{II}, adiabatic efficiency η and polytropic efficiency η_∞ (see Sect.1.7.3) is expressed graphically in Fig. C.3 and in Fig. C.4, which refer to the air compression for compression pressure ratios from 2 to 10 for initial temperatures of 30 °C and 300 °C and to the air expansion with the same pressure ratio values and initial expansion temperatures of 550 °C and 950 °C, respectively.

Regarding the compression, η_{II} is always greater than $\eta_{C,\infty}$ and grows for the high compression values, since the dissipations produce their effects at higher

Fig. C.5 Second-law efficiency as a function of the recuperative regenerator efficiency for a gas-turbine cycle

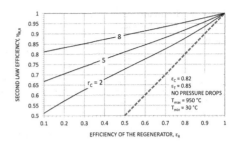

temperatures. Likewise, in the expansions starting from a high initial temperature, η_{II} is always very high.

As far as the significance of the different efficiencies, η_∞ is the most reliable index of the fluid dynamic quality of a compression or expansion process; η permits the immediate formulation of an overall energy balance; while η_{II} ought to be the most significant parameter of merit at a thermodynamic level, in reality it has connotations that are a little too theoretical, given that its definition implicitly hypothesises a perfect use (reversible) of the internal energy increase, due to the dissipations.

The behavior of η_{II} for a heat exchanger is exemplified in Fig. C.5 with reference to the regenerator of a gas-turbine cycle. As the compression value increases, the quantity of heat that is possible to regenerate diminishes and the average temperature at which the heat exchange takes place increases; both factors contribute to a substantial increase of η_{II}.

C.4 The Losses in Steam Cycles and in Gas-Turbine Cycles

(a) Reference Cycles

With the aim of establishing a method of analysis for the losses which will give better understanding of the thermodynamic phenomenon and will suggest more profitable approaches for improving efficiency, we will examine two cycles that represent almost all the thermal systems for generating electricity: the steam cycle and the gas-turbine cycle.

As our basic reference, we shall consider a steam cycle with a simple regeneration and a gas cycle which is regenerative, too (v. la Fig. C.6). The parameters of merit for the components, where not explicitly made to vary, will follow the values reported in Fig. C.6.

Based on (C.3), in the case of the Rankine cycle, the efficiency losses as a function of the vaporisation temperature followed the trend illustrated in Fig. C.7. Although an isothermal source normally replaces the flow of the high-temperature combustion gases at the highest temperature in the cycle, most of the loss is linked to the irreversibility of the heat exchange in the furnace (for example, 60 % of the

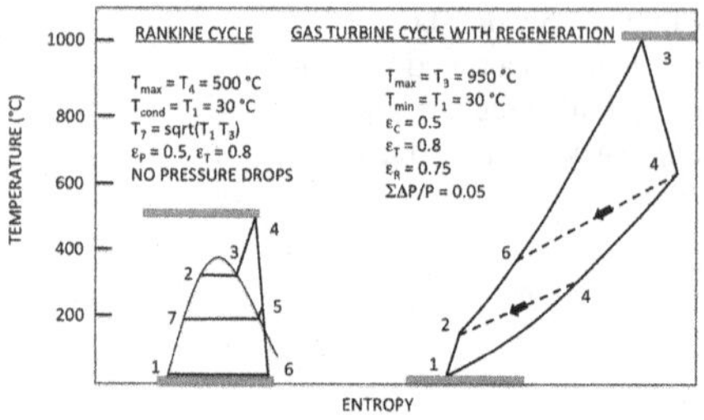

Fig. C.6 Steam and gas
reference cycles for loss
analysis

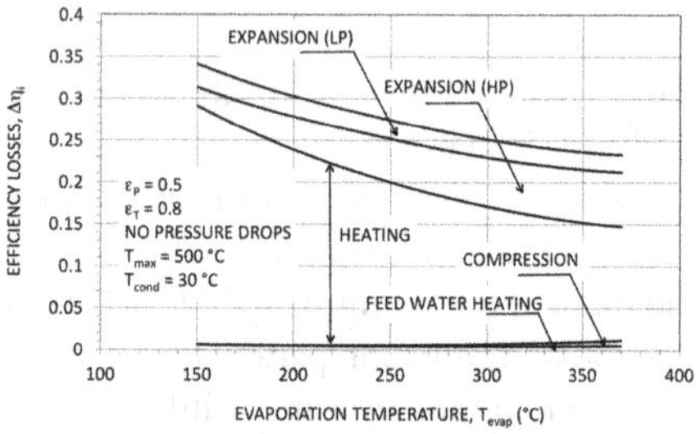

Fig. C.7 Efficiency losses
for the steam cycle as a result
of the entropy analysis

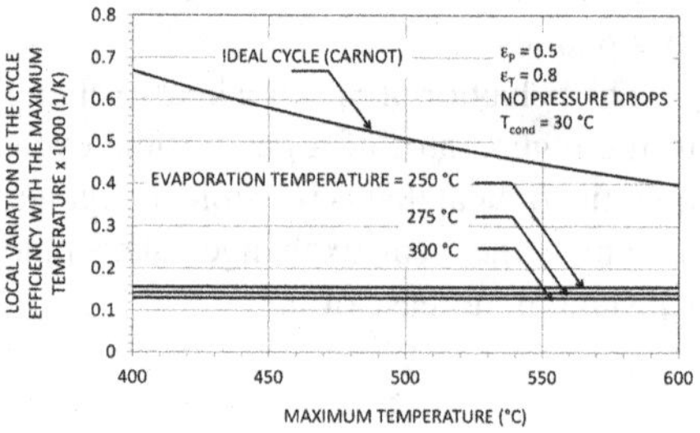

Fig. C.8 Local dependence
of Rankine cycle performance
on maximum temperature

total). The irreversibilities in the turbine and those, small ones, in the feedwater heater account for almost all the remainder of the losses (for example, 30 % and 7 % of the total, respectively). At the extreme temperatures of 500 and 30 °C, the efficiency for the maximum pressure of 80 bar is equal to 60 % of that of the Carnot cycle (second-law efficiency).

The sensitivity of the efficiency to variations in the maximum temperature $((\partial\eta/\partial T_{max})$; see Fig. C.8) is, on average, just 1/4 of that of the Carnot cycle. By contrast, with regard to the minimum temperature, the sensitivity is practically identical to that of the ideal cycle $((\partial\eta/\partial T_{cond})$; see Fig. C.9). These results explain the reasons (at least, till now) of the low interest to develop materials that can permit higher temperatures at turbine entry, whilst every care has always been placed in the design of an efficient system for cooling the condenser.

In the case of the regenerative gas cycle, the main contributor to losses is the release of waste heat into the environment (about 50 % of the total for $r_C = 5$), followed, in equal measure, by the fluid dynamic irreversibilities in the compressor and in the turbine and the loss through heating at high temperature. The loss in the regenerator is greatly affected by the compression level and is slight evanescent

Fig. C.9 Local dependence
of Rankine cycle performance
on minimum temperature

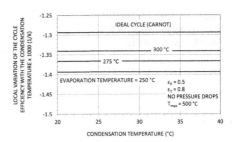

Fig. C.10 Distribution of the
losses, on the basis of
entropic analysis, in a
regenerative turbine gas cycle

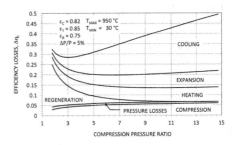

Fig. C.11 Local sensitivity
of a regenerative gas cycle to
the maximum temperature

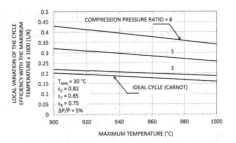

for $r_C > 7 - 8$ (see Fig. C.10). Under optimum conditions, the cycle efficiency is approximately equal to 45 % of that of the Carnot cycle. The sensitivity of the cycle to the maximum temperature essentially depends on the compression value. At around $T_{max} = 950\,°C$, it is equal to that of the Carnot cycle for $r_C = 3$, being 1.5 times greater for $r_C = 5$, double for $r_C = 8$ (Fig. C.11). The sensitivity of the cycle with regard to T_{min} (Fig. C.12) shows a similar trend, with the same progression as a function of r_C, with absolute numerical values about three times those of Fig. C.11. Doing without the regeneration (simple gas cycle), the distribution of the losses undergoes a significant alteration with a notable increase in the irreversibilities of the heat exchange, especially at low temperature, accompanied by a big reduction in efficiency losses caused by turbomachines especially at low temperature (a reduction caused by the increased heat consumption; Fig. C.13).

Fig. C.12 Local sensibility
of a regenerative gas cycle to
the minimum temperature

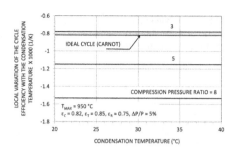

Fig. C.13 Distribution of the
losses, on the basis of
entropic analysis, in a turbine
gas cycle without
regeneration (simple cycle)

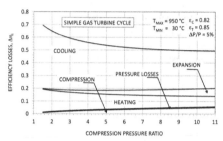

C.5   Significance of the Thermodynamic Losses and Investigation of the Indirect Effects of the Irreversibilities

The results of Fig. C.13 may seem surprising in so far as they show small losses
for the turbomachines (for example, 15 % of the total when adding those of the
compressor to those of the turbine) which appears contrary to the common opinion
that these losses are directly responsible for the poor performance of the gas cycles
(22.4 % per r_C = 5). The high efficiency of the second principle $\eta_{C,II}$ of the
compressor and, especially, of the turbine, for which there is no correspondingly
high energy efficiency—if highlighted by other means—prompts the suspicion that
the representation in Fig. C.13 hides the effective responsibility of the inefficiencies
of the various components. In order to shed light on this final circumstance, we
can proceed in the following way. For example, eliminating the irreversibility of the
turbine, hypothesising a unitary expansion efficiency and comparing the increased
cycle efficiency that results with the entity of the losses in the turbine that are
directly removed (calculated by means of the production of eliminated entropy). In
the absence of indirect effects, the two quantities should be identical and their ratio
should be equal to one. However, as illustrated in Fig. C.14, the ratio results much
higher than one in the case of the turbine (even higher than 3) and a little more than
one in the case of the compressor. Removing the loss in the turbine significantly
reduces the irreversibility of the heat exchange in releasing the waste heat into
the atmosphere, and this second effect frequently has a greater final influence on
efficiency than the direct influence. Towards the higher compression values, the
indirect effect of the losses diminishes, but the direct effect grows (that is, the

Fig. C.14 Ratio between the effective losses and the entropic losses in a simple turbine gas cycle

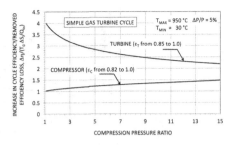

production of entropy in expansion and compression) and the cycle efficiency does not improve. If the presence of the regenerator reduces the indirect effects of the losses in the turbine (the added heat discharged by the turbine is used relatively efficiently by the regenerator), it increases the indirect losses of the compressor (the increase in entropy and the temperature at the end of compression produces a parallel increase in the temperature of the gases discharged into the atmosphere at the end of regeneration; see Fig. C.15).

No less important is the multiplication effect of removing the loss in the regenerator, obtained by hypothesising a unitary regeneration efficiency. In fact, by rendering the heat exchange in the regenerator reversible, we render less irreversible both the introduction of heat (a higher temperature at the start of heating) and the release of the waste heat (a lower temperature at the start of releasing heat into the atmosphere). The sharp rise in the ratio "useful global effect/loss removed" with the growth in the compression level is justified by the higher cost of the indirect losses (due to the fall in the temperature at the start of heating and the increase in the temperature at the start of cooling) as the compression level rises.

If, using the same criterion, the ratio defined above was sought for the turbine or pump losses in a Rankine cycle, then we would get essentially unitary values, signifying the absence of important indirect effects of irreversibility.

The global effect reported in Fig. C.15 can be broken down into the constituent elements that represent the influence on the loss $\Delta\eta_i$ of each component by the removal of a single loss (first in the turbine, then the compressor and finally the regenerator). Table C.1 analyzes in detail the calculation results for a compression rate $r_C = 5$. It can be seen, for example, how the removal of the loss in the regenerator has a useful multiplication effect with a factor of 2.45 ($= 0.0908/0.0371$). The greatest contribution to increased efficiency comes from reducing the loss in cooling (0.0671); this is followed by the direct effect (0.0371) and that from reducing the loss in heating (0.0146). There are also small negative effects (compressor -0.0127; expander -0.0129) linked to the drop in the heat being introduced into the cycle.

Wishing to examine a broader range of variations in the component performances, we could calculate the distribution of the losses as a function of the parameter of merit of the compressor, turbine, and regenerator. The results, shown, by way of example, in a graph in Figs. C.16 and C.17, naturally prompt several observations. It is worth noting the influence of the regeneration efficiency in a low-level compression cycle ($r_C = 3$, great weight of the regenerator on the

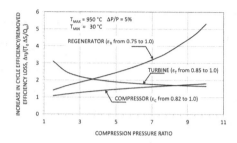

Fig. C.15 Ratio between the effective losses and the entropic losses in a regenerative turbine gas cycle

Table C.1 Analysis of the losses in regenerative turbine gas cycles ($r_C = 5$)

Component	Reference cycle $\Delta\eta_{i,rif}$	$\epsilon_T = 1.0$ $\Delta\eta_i$	$\dfrac{\Delta\eta_{i,ref}}{-\Delta\eta_i}$ $\epsilon_T = 1.0$ $\Delta\eta_i$	$\epsilon_C = 1.0$ $\Delta\eta_i$	$\dfrac{\Delta\eta_{i,ref}}{-\Delta\eta_i}$ $\epsilon_C = 1.0$ $\Delta\eta_i$	$\epsilon_R = 1.0$ $\Delta\eta_i$	$\dfrac{\Delta\eta_{i,ref}}{-\Delta\eta_i}$ $\epsilon_R = 1.0$
Compressor	0.0489	0.0442	0.0047	0.0000	**0.0489**	0.0616	−0.0127
Pressure losses	0.0092	0.0083	0.0009	0.0090	0.0002	0.0116	−0.0024
Regenerator	0.0371	0.0253	0.0118	0.0470	−0.0099	0.0000	**0.0371**
Heating	0.0559	0.0642	−0.0083	0.0575	−0.0016	0.0413	0.0146
Expansion	0.0494	0.0000	**0.0494**	0.0485	0.0009	0.0623	−0.0129
Cooling	0.2036	0.1685	0.0351	0.1706	0.0330	0.1365	0.0671
Total	0.4041	0.3105	0.0936	0.3325	0.0715	0.3134	0.0908
η_{cycle}	0.3481	0.4417		0.4197		0.4388	

cycle efficiency). With ϵ_R varying from zero to one, the direct loss takes on its maximum value for intermediate performances (0.6–0.7) and becomes nullified at both extremes of the variability range. This circumstance should remind us that a modest loss in a component does not necessarily mean a high-quality component, since its quantitative effect on the energy process plays a major role in determining the loss. Likewise, we can see the different role played by the essential components (compressor, turbine) and those that are accessories (regenerator). Furthermore, Fig. C.16 highlights the expansion of the losses for heating and cooling, induced by the reduction in the regenerator's parameter of merit. The opposite effect can be noted for the turbomachines. Similarly, there is a clear distinction between those components whose energy performances are directly controlled by the designer (turbine, compressor, and regenerator) and those whose performances are defined indirectly (heater and cooler).

Similar considerations but referring to an essential component (the compressor) and at a compression level making it quantitatively important ($r_C = 8$) are illustrated in Fig. C.17. What is particularly significant is the nullifying of the cycle efficiency ($\sum \Delta\eta_i \approx 0.7$) with an adiabatic compression efficiency of around 0.55.

Fig. C.16 Entropy losses as a function of the regenerator efficiency in a gas cycle

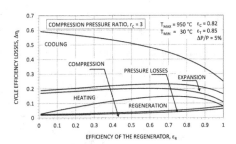

Fig. C.17 Entropy losses as a function of the compression efficiency in a gas cycle

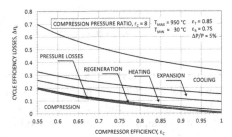

C.6   Alternative Methods for Evaluating the Influence of Losses

We have seen how entropic analysis in its normal form, associating each component with a loss that is proportional to the production of entropy that takes place within it, does not clarify the causal nexus between the component performances and the final energy result. However, that nexus represents the most efficient instrument for the designer when evaluating the feasibility of investing resources in the improvement of the energy characteristics of the various components.

So, as an alternative to measuring the effect of the irreversibility by means of the production of entropy, the importance of the energy quality of a component can be expressed by a coefficient of influence defined as follows:

$$K_i = \frac{\partial \eta / \eta}{\partial \epsilon_i / \epsilon_i} = \frac{\partial \eta}{\partial \epsilon_i} \frac{\epsilon_i}{\eta} \tag{C.8}$$

in which, η is the cycle efficiency and ϵ_i is the parameter of merit (conventional) of the generic component and which gives the relative increase in the final efficiency of the cycle in relation to the relative increase in the component's parameter of merit. One example of the kind of information that can be obtained in this way is shown in Fig. C.18 which records the behavior of K_i according to variation in the quality factor of the compressor, turbine and regenerator for $r_C = 3$ and $r_C = 5$ for regenerative cycles and of the compressor and turbine for simple cycles ($r_C = 5$).

In all cases, it is the turbine which is the component with the greatest influence (K_i always greater than two for the simple cycle and equal to around 1.3 in the

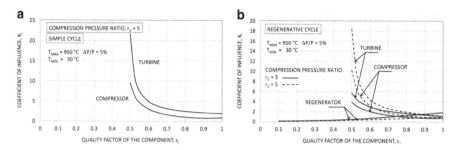

Fig. C.18 Influence of the component performances on the efficiency of gas cycles. (**a**) Simple cycle. (**b**) Cycle with regeneration and two different levels of compression

regeneration cycles in question). The importance of the compressor is significantly lower (values about one in the zone of the diagrams in question). For cycles with a low compression ratio ($r_C = 3$), the coefficient of influence of the regenerator is also about one, a value which is halved for $r_C = 5$. It is also worth noting the sharp rise in K_i values for the turbomachines as their efficiency diminishes.

If we calculate similar coefficients for the Rankine cycle, with reference to the turbine and the pump, we get substantially unitary values for the turbine and very small ones for the pump. The overall picture confirms the critical nature of the turbomachine efficiencies in gas cycles.

C.7   Final Considerations

Such an analysis of the losses gives a greater understanding of the causes and circumstances that determine the energy performance of the power cycles. Below, we give several examples of interpreting the way in which some of the thermodynamic cycles operate.

(a) Rankine Steam Cycle

The entropic analysis gives a realistic distribution of the losses. The fact that the most important loss is constituted by the heat exchange irreversibilities at high temperatures suggests the main guideline to improving the cycle efficiency: an increase in the turbine entry pressure, even to super-critical values, or the use of binary or combined cycles.

(b) Simple Gas Cycle

The entropic analysis underestimates the importance of the losses in the turbomachines. Referring to the diagrams in Fig. C.19a, in correspondence with the optimum compression level, we see that the loss connected with the cooling of the discharged

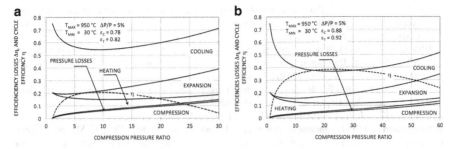

Fig. C.19 Energy optimisation of simple gas cycles

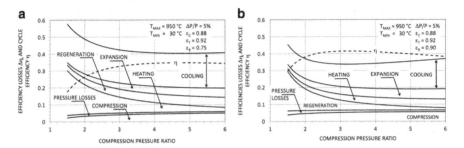

Fig. C.20 Energy optimisation of regenerative gas cycles

gases is preponderant. The losses in the machines, which are already significant, increase considerably if the indirect effects are taken into consideration. Their rapid increase with r prevents the attainment of those cycle configurations that would correspond to moderate losses for the cooling and heating. That objective may be reached, though, by improved performances in the machines (Fig. C.19b).

(c) Regenerative Gas Cycle

Similar considerations to those for the simple cycle, although mitigated by the lower compression levels involved, also apply here for the role of the losses in the turbomachines. In this case, too, the loss through discharging heat at the cold source is predominant and could be reduced by using lower compression levels, which, however, are only compatible with very high regeneration efficiencies (see Fig. C.20a, b).

References

1. Trimble LC, Owens WL (1980) Review of mini-OTEC performance. In: Proceedings of the 15th Intersociety Energy Conversion Engineering Conference, Seattle, pp 1331–1338
2. Johnson FA (1992) Closed cycle ocean thermal energy conversion. In: Seymour RJ (ed) Ocean energy recovery: the state of the art. American Society of Civil Engineers, New York

3. Gauthier M, Marvaldi J, Zangrando F (1992) State of the art in open-cycle ocean thermal energy conversion. In: Seymour RJ (ed) Ocean energy recovery: the state of the art. American Society of Civil Engineers, New York
4. Bharathan D (2011) Staging Rankine cycles using ammonia for OTEC power production. NREL - National Renewable Energy Laboratory. Technical Report NREL/TP-5500-49121. Contract No. DE-AC36-08GO28308. Golden, Colorado
5. Anon (1987) Progetto Finalizzato Energetica. Tema Pompe di Calore. Consiglio Nazionale delle Ricerche. Technical Report della Soc, Aerimpianti
6. Anon (1957) L'Italia fisica. Chapter VI. Touring Club Italiano, Milano
7. Santomauro L (1957) Lineamenti climatici di Milano. Quaderni della Citta' di Milano. Milano
8. Fleury J, Bellot Ch (1984) Ammonia bottoming cycle development at Electricité de France for nuclear power plants. In: ORC–HP–technology. working fluids problems. Proceedings of the International VDI-Seminar. VDI Berichte 539. VDI Verlag, Düsseldorf